BOOK 2

NAME:

CLASS:

NELSON MATHS

WORKBOOK

YEARS 7–10

Kuldip Khehra
Judy Binns
Gaspare Carrozza
Robert Yen
Sandra Tisdell-Clifford

PUZZLE SHEETS

Nelson Maths Workbook 2
1st Edition
Kuldip Khehra
Judy Binns
Gaspare Carrozza
Robert Yen
Sandra Tisdell-Clifford
ISBN 9780170454520

Publisher: Robert Yen
Project editor: Alan Stewart
Editor: Anna Pang
Cover design: James Steer
Original text design by Alba Design, Adapted by: James Steer
Project designer: James Steer
Permissions researcher: Corrina Gilbert
Typeset by: MPS Limited
Production controller: Karen Young
Text illustrations: Cat MacInnes

Any URLs contained in this publication were checked for currency during the production process. Note, however, that the publisher cannot vouch for the ongoing currency of URLs.

For product information and technology assistance,
in Australia call **1300 790 853**;
in New Zealand call **0800 449 725**

For permission to use material from this text or product, please email **aust.permissions@cengage.com**

ISBN 978 0 17 045452 0

Cengage Learning Australia
Level 7, 80 Dorcas Street
South Melbourne, Victoria Australia 3205

Cengage Learning New Zealand
Unit 4B Rosedale Office Park
331 Rosedale Road, Albany, North Shore 0632, NZ

For learning solutions, visit **cengage.com.au**

Printed in China by 1010 Printing International Limited.
1 2 3 4 5 6 7 24 23 22 21 20

PREFACE

This 200-page workbook contains worksheets, puzzles, StartUp assignments and homework assignments written for the Australian Curriculum in Mathematics. It can be used as a valuable resource for teaching Year 8 mathematics, regardless of the textbook used in the classroom, and takes a wholistic approach to the curriculum, including some Year 7 and Year 9 work as well. This workbook is designed to be handy for homework, assessment, practice, revision, relief classes or 'catch-up' lessons.

Inside:

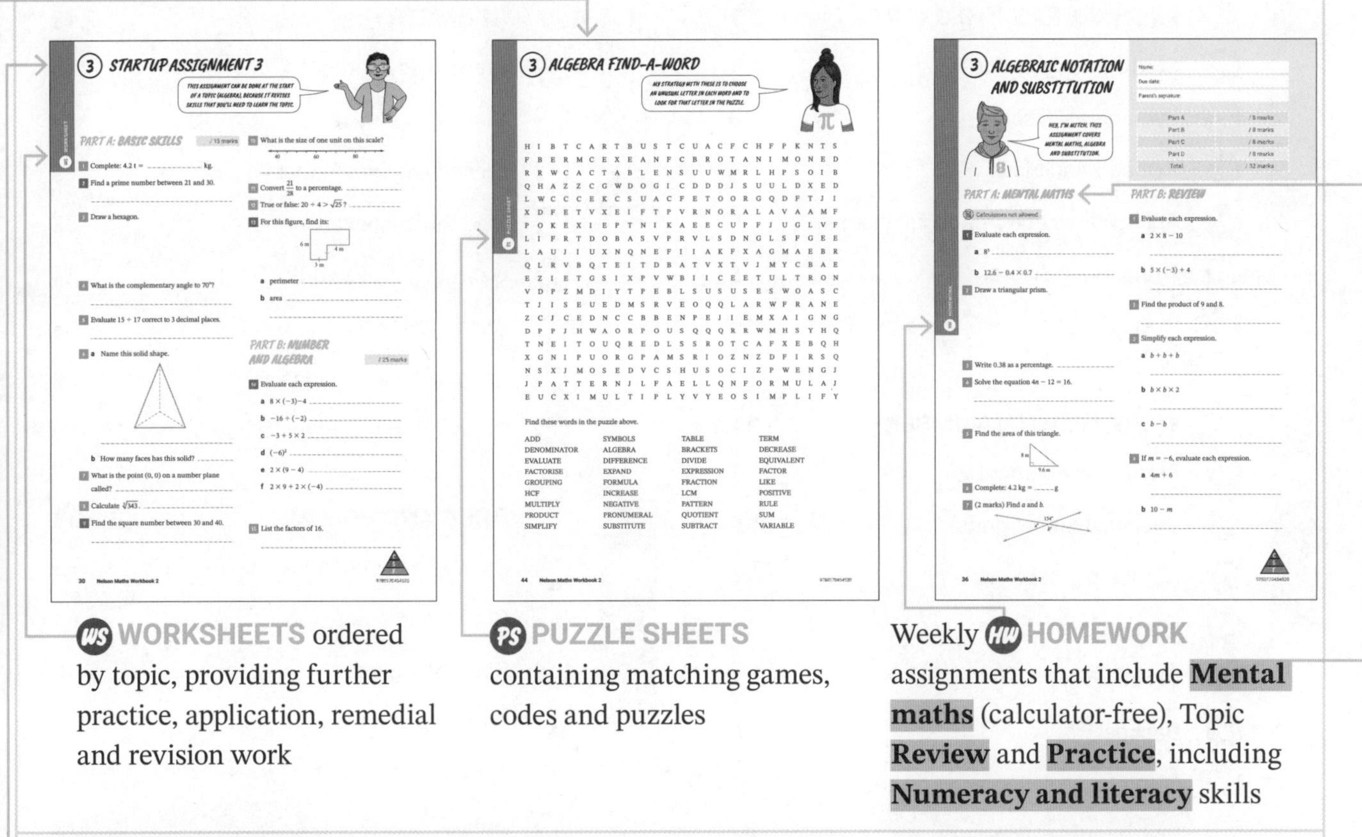

WS **WORKSHEETS** ordered by topic, providing further practice, application, remedial and revision work

PS **PUZZLE SHEETS** containing matching games, codes and puzzles

Weekly **HW** **HOMEWORK** assignments that include **Mental maths** (calculator-free), Topic **Review** and **Practice**, including **Numeracy and literacy** skills

StartUp assignments beginning each topic, revising skills from previous topics and prerequisite knowledge for the topic, including basic skills, review of a specific topic and a challenge problem

Word puzzles, such as a crossword or find-a-word, that reinforce the language of mathematics learned in the topic

The ideas and activities presented in this book were written by practising teachers and used successfully in the classroom.

Colour-coding of selected questions

Questions on most worksheets are graded by level of difficulty:

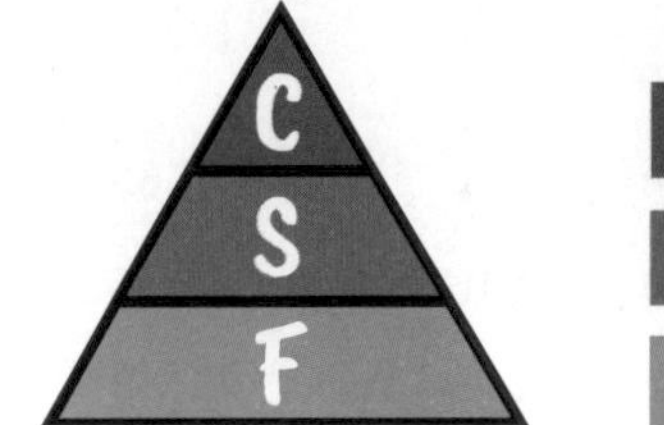

Complex

Standard

Foundation

CONTENTS

WORKSHEET

PUZZLE SHEET

HOMEWORK

CONTENTS

MEET YOUR MATHS GUIDES ...

INTRODUCING MS LEE.

THIS WORKBOOK CONTAINS WORKSHEETS, PUZZLE SHEETS AND HOMEWORK ASSIGNMENTS

HI, I'VE BEEN TEACHING MATHS FOR OVER 20 YEARS

I BECAME GOOD AT MATHS THROUGH PRACTICE AND EFFORT

I WILL GUIDE YOU THROUGH THE WORKSHEETS

MATHS IS ABOUT MASTERING A COLLECTION OF SKILLS, AND I CAN HELP YOU DO THIS

THIS IS ZINA, A MATHS TUTOR AND MS LEE'S YEAR 12 STUDENT

HEY, I LOVE CREATING AND SOLVING PUZZLES

NOT JUST MATHS PUZZLES BUT WORD PUZZLES TOO!

PUZZLES HELP YOU THINK IN NEW AND DIFFERENT WAYS

LET ME SHOW YOU HOW, AND YOU'LL GET SMARTER ALONG THE WAY

9780170454520

SAY HI TO MITCH, ZINA'S FRIEND AND A STAR BASKETBALLER
MITCH DOES A LOT OF TRAINING FOR HIS SPORT
HE'S ALWAYS ON THE COURT PRACTISING HIS SHOTS
HEY, I'M YOUR HOMEWORK GUIDE
I'LL TRAIN YOU ON A PROGRAM OF DIFFERENT DRILLS
I'LL MAKE YOU A MATHS ATHLETE ... A MATHLETE!

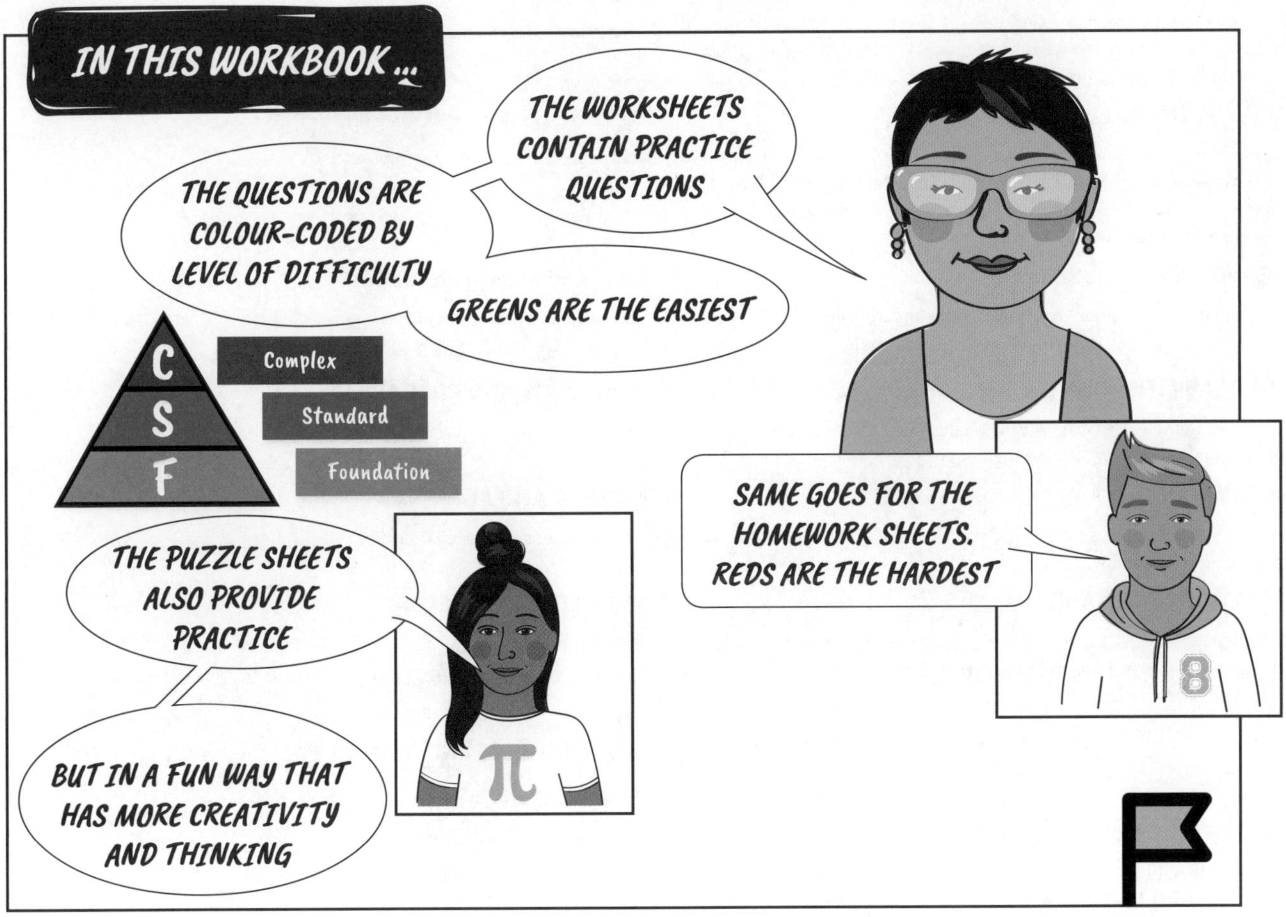

IN THIS WORKBOOK ...

THE WORKSHEETS CONTAIN PRACTICE QUESTIONS

THE QUESTIONS ARE COLOUR-CODED BY LEVEL OF DIFFICULTY

GREENS ARE THE EASIEST

C Complex

S Standard

F Foundation

SAME GOES FOR THE HOMEWORK SHEETS. REDS ARE THE HARDEST

THE PUZZLE SHEETS ALSO PROVIDE PRACTICE

BUT IN A FUN WAY THAT HAS MORE CREATIVITY AND THINKING

SO LET'S GET GOING.

GAME ON!

CURRICULUM GRID

Chapter and content	Australian curriculum strand and substrand
1 PYTHAGORAS' THEOREM	**MEASUREMENT AND GEOMETRY**
Square roots, Pythagoras' theorem, Testing for right-angled triangles, Pythagorean triads, Pythagoras' theorem problems	Pythagoras and trigonometry
2 WORKING WITH NUMBERS	**NUMBER AND ALGEBRA**
Mental calculation, Operations with integers, Order of operations, Decimals, Operations with decimals, Terminating and recurring decimals, Powers and roots, Prime factors, Index laws	Number and place value Real numbers
3 ALGEBRA	**NUMBER AND ALGEBRA**
Variables, From words to algebraic expressions, Substitution, Simplifying algebraic expressions, Expanding expressions, Factorising expressions	Patterns and algebra
4 GEOMETRY	**MEASUREMENT AND GEOMETRY**
Angle geometry, Line and rotational symmetry, Classifying triangles and quadrilaterals, Properties of quadrilaterals, Angle sums of triangles and quadrilaterals	Location and transformation Geometric reasoning
5 AREA AND VOLUME	**MEASUREMENT AND GEOMETRY**
Perimeter, Metric units for area, Areas of rectangles, triangles, parallelograms and composite shapes, Areas trapeziums, kites and rhombuses, Parts of a circle, Circumference and area of a circle, Perimeter and area of a sector, Metric units for volume, Volumes of prisms and cylinders, Volume and capacity	Using units of measurement
6 FRACTIONS AND PERCENTAGES	**NUMBER AND ALGEBRA**
Fractions, Operations with fractions, Percentages, fractions and decimals, Fraction and percentage of a quantity, Expressing amounts as fractions and percentages, Percentage increase and decrease, The unitary method, Profit, loss and GST, Percentage problems	Real numbers Money and financial mathematics
7 INVESTIGATING DATA	**STATISTICS AND PROBABILITY**
Organising and displaying data, Types of data, The mean, mode, median and range, Analysing frequency tables, Dot plots, Stem-and-leaf plots, Frequency histograms and polygons, Sampling, Designing survey questions, Comparing samples and populations	Data representation and interpretation
8 CONGRUENT FIGURES	**MEASUREMENT AND GEOMETRY**
Transformations, Congruent figures, Constructing triangles, Tests for congruent triangles, Proving properties of quadrilaterals	Location and transformation Geometric reasoning
9 PROBABILITY	**STATISTICS AND PROBABILITY**
Probability, Complementary events, Venn diagrams, Two-way tables, Experimental probability, Relative frequency	Chance
10 EQUATIONS	**NUMBER AND ALGEBRA**
Two-step equations, Equations with variables on both sides, Equations with brackets, Equation problems	Linear and non-linear relationships
11 RATIOS, RATES AND TIME	**NUMBER AND ALGEBRA**
Ratios, Ratio problems, Scale maps and plans, Dividing a quantity in a given ratio, Rates, Best buys, Rate problems, Speed, Travel graphs, Time differences, World time zones	Real numbers Money and financial mathematics Linear and non-linear relationships
	MEASUREMENT AND GEOMETRY
	Using units of measurement
12 GRAPHING LINEAR EQUATIONS	**NUMBER AND ALGEBRA**
Tables of values, Finding the rule, The number plane, Graphing number patterns, Graphing linear equations, Finding the equation of a line, Solving linear equations graphically, Intersecting lines	Linear and non-linear relationships

STARTUP ASSIGNMENT 1

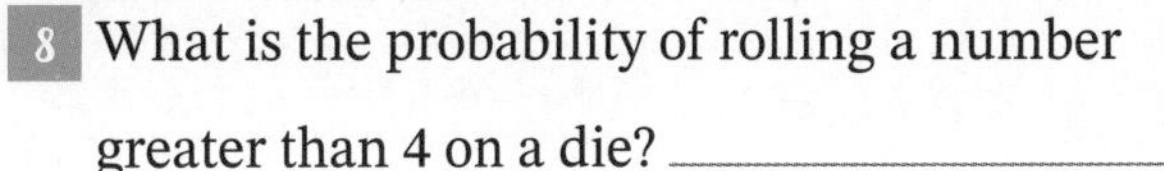

HERE ARE SOME SKILLS YOU NEED TO KNOW
TO DO WELL IN MATHS. PART A IS MIXED
SKILLS, PART B IS FOR THE PYTHAGORAS'
THEOREM TOPIC WE'RE STARTING.

WS WORKSHEET

PART A: BASIC SKILLS

/ 15 marks

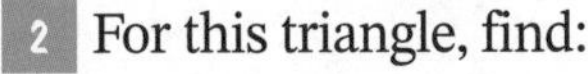

1 Of the 200 students at the school disco, 36% are male. How many is this?

____

2 For this triangle, find:

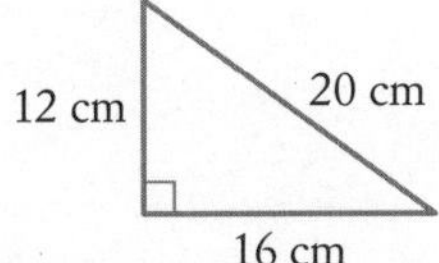

a its perimeter ______________________

b its area ______________________

3 If a coin is tossed, what is the sample space?

____

4 Simplify each expression.

a $m + m + 2m$ ______________________

b $4 \times h \times 2$ ______________________

5 What is the supplementary angle to 45°?

____

6 Solve each equation.

a $x + 8 = -1$

____

b $2d - 5 = 9$

____

7 Convert $\frac{17}{40}$ to a percentage.

____

8 What is the probability of rolling a number greater than 4 on a die? ______________________

9 If $p = -3$, evaluate $4p + 7$.

____

10 Find $\frac{3}{8}$ of 5 kg in grams.

____

11 Simplify 18 : 8. ______________________

12 Evaluate 0.428×1000. ______________________

PART B: POWERS AND ROOTS

/ 25 marks

13 Evaluate each expression.

a $7^2 - 3^2$ ______________________

b $8^2 + 15^2$ ______________________

c 9^3 ______________________

d 3.2^2 ______________________

e 4^4 ______________________

f $\sqrt{441}$ ______________________

g $\sqrt{10^2 + 24^2}$ ______________________

h $\sqrt[3]{216}$ ______________________

14 (2 marks) Draw an isosceles right-angled triangle and label the longest side x.

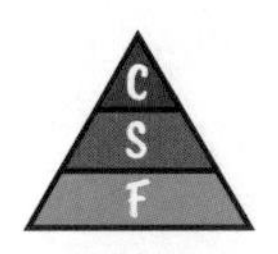

WS WORKSHEET

15 True or false?

a $9^2 + 12^2 = 15^2$ ______

b $8^2 + 10^2 = 14^2$ ______

c $8^2 - 3^2 = 5^2$ ______

d $13^2 - 12^2 = 5^2$ ______

e $\sqrt{16+4} = \sqrt{16} + \sqrt{4}$ ______

f $\sqrt{16\times4} = \sqrt{16} \times \sqrt{4}$ ______

16 (3 marks) Circle the 3 square numbers from this list:

49, 200, 225, 81, 46, 80, 99, 65.

17 Evaluate correct to 2 decimal places:

a $\sqrt{24}$ ______

b $\sqrt{12^2 - 6^2}$ ______

c $\sqrt{40}$ ______

d $\sqrt[3]{160}$ ______

18 (2 marks) Construct this triangle with lengths $AB = 4$ cm, $BC = 3$ cm and measure the length of AC.

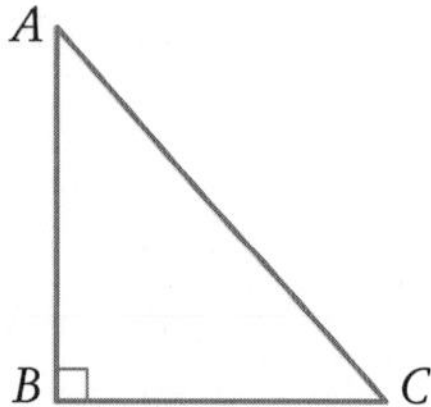

PART C: CHALLENGE

Bonus / 3 marks

Draw each figure without lifting your pen.

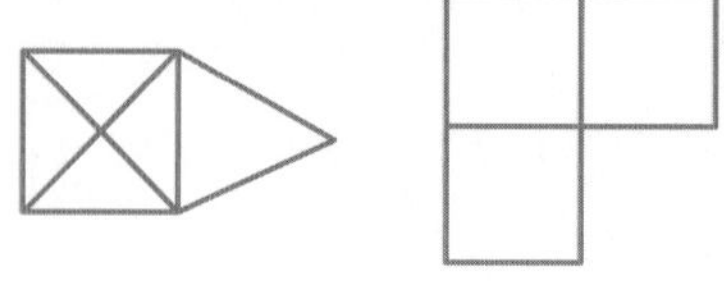

C
S
F

9780170454520

PYTHAGORAS' DISCOVERY 1

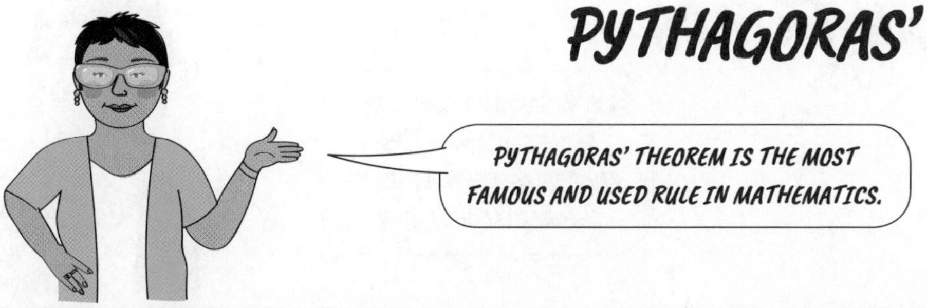

PYTHAGORAS' THEOREM IS THE MOST FAMOUS AND USED RULE IN MATHEMATICS.

WS WORKSHEET

Teacher's tickbox

For each triangle:

- ❑ circle the hypotenuse
- ❑ write Pythagoras' theorem
- ❑ square the length of each side
- ❑ show that Pythagoras' theorem is true.

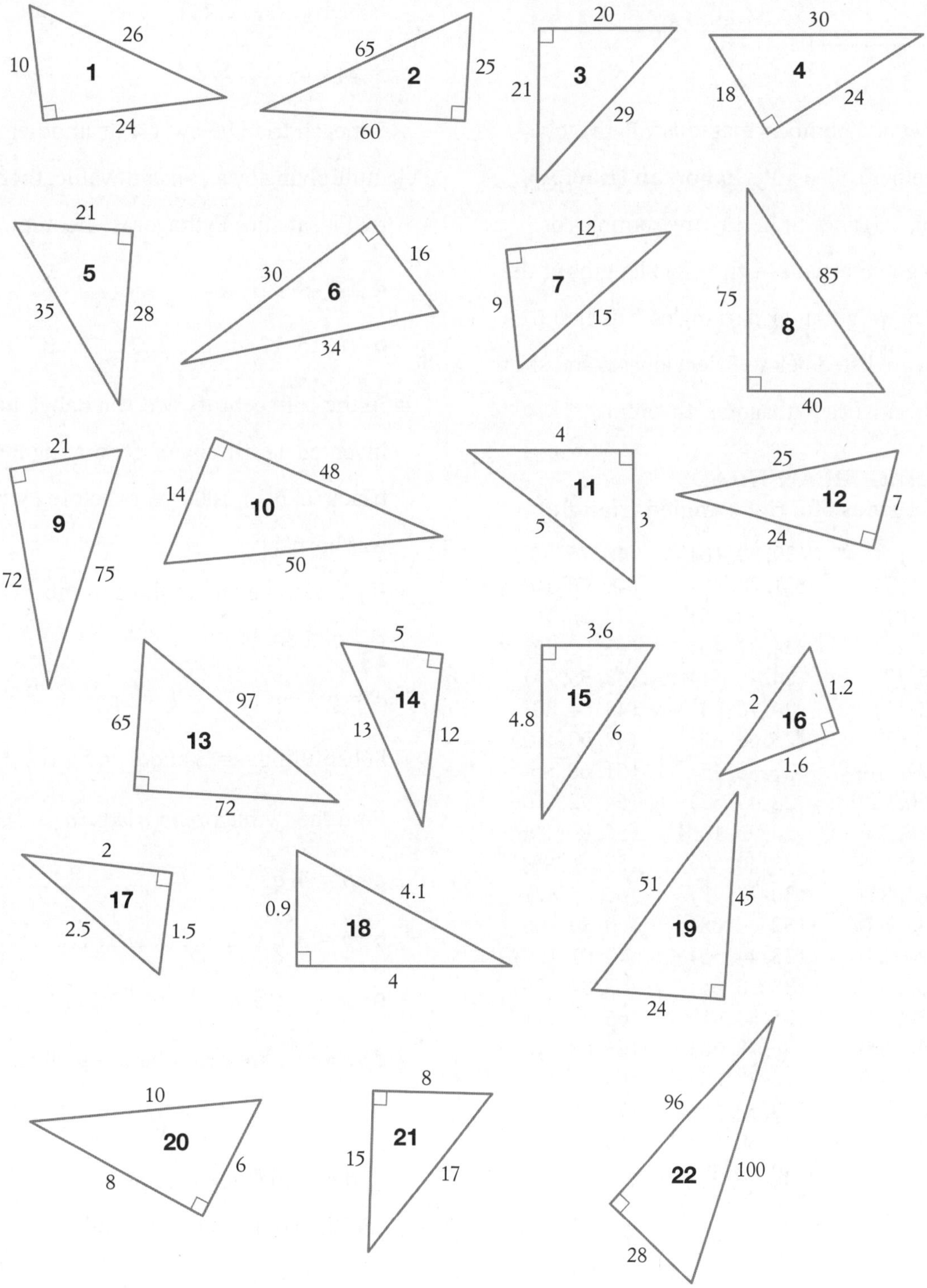

1 PYTHAGOREAN TRIADS

PYTHAGOREAN TRIADS ARE GROUPS OF 3 NUMBERS THAT FOLLOW PYTHAGORAS' THEOREM, SUCH {3, 4, 5}.

1 Construct each triangle accurately and measure the length of the **hypotenuse**.
Verify that they satisfy Pythagoras' theorem.

a

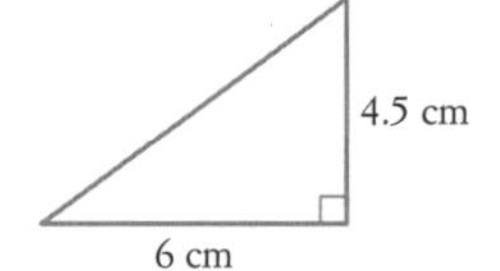

b

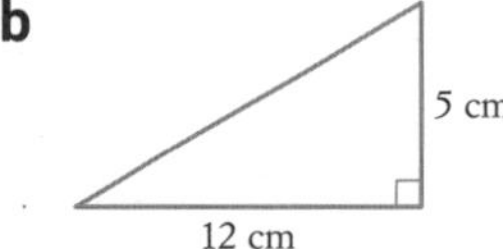

2 Any set of 3 numbers that satisfy Pythagoras' theorem is called a **Pythagorean triad.** So (4.5, 6, 7.5) and (5, 12, 13) are examples of Pythagorean triads — the exact lengths of the sides of a right-angled triangles. The first 62 triads are listed below. Select any set and show that it satisfies Pythagoras' theorem.

PYTHAGOREAN TRIADS
(exact sides of a right-angled triangle)

(3, 4, 5)	(20, 99, 101)	(40, 75, 85)
(5, 12, 13)	(21, 28, 35)	(40, 96, 104)
(6, 8, 10)	(21, 72, 75)	(42, 56, 70)
(7, 24, 25)	(24, 32, 40)	(45, 60, 75)
(8, 15, 17)	(24, 45, 51)	(48, 55, 73)
(9, 12, 15)	(24, 70, 74)	(48, 64, 80)
(9, 40, 41)	(25, 60, 65)	(48, 90, 102)
(10, 24, 26)	(27, 36, 45)	(51, 68, 85)
(11, 60, 61)	(28, 45, 53)	(54, 72, 90)
(12, 16, 20)	(28, 96, 100)	(56, 90, 106)
(12, 35, 37)	(30, 40, 50)	(57, 76, 95)
(13, 84, 85)	(30, 72, 78)	(60, 63, 87)
(14, 48, 50)	(32, 60, 68)	(60, 80, 100)
(15, 20, 25)	(33, 44, 55)	(60, 91, 109)
(15, 36, 39)	(33, 56, 65)	(63, 84, 105)
(16, 30, 34)	(35, 84, 91)	(65, 72, 97)
(16, 63, 65)	(36, 48, 60)	(66, 88, 110)
(18, 24, 30)	(36, 77, 85)	(69, 92, 115)
(18, 80, 82)	(39, 52, 65)	(72, 96, 120)
(20, 21, 29)	(39, 80, 89)	(80, 84, 116)
(20, 48, 52)	(40, 42, 58)	

3 Multiplying (or dividing) each number in a triad by the same value gives another triad. For example:

(3, 4, 5) × 2 gives (6, 8, 10)

× 3 gives (9, 12, 15)

÷ 2 gives (1.5, 2, 2.5)

For each triad below, create another triad by multiplying by a constant value, then check that it satisfies Pythagoras' theorem.

a (7, 24, 25) ____________________

b (8, 15, 17) ____________________

4 In the 16th century BCE, the Babylonians invented a formula for generating Pythagorean triads (a, b, c), 1000 years before Pythagoras was born!

If p and q are any 2 whole numbers and $p > q$, then:

$a = p^2 - q^2$ $\quad$ $b = 2pq$ $\quad$ $c = p^2 + q^2$

Substituting $p = 3$ and $q = 2$ gives (5, 12, 13).

Find the Pythagorean triads (a, b, c) if:

a $p = 4, q = 3$ ____________________

b $p = 3, q = 1$ ____________________

c $p = 6, q = 4$ ____________________

d p and q are 2 numbers you choose.

____

(Advanced students: notice that a^2 is a square multiple of the sum of b and c.)

9780170454520

5 Pythagoras himself invented this formula for generating Pythagorean triads.

If n is any odd number, then:

$a = n, b = \frac{n^2 - 1}{2}, c = \frac{n^2 + 1}{2}$.

Substituting $n = 3$ gives (3, 4, 5).

Find the Pythagorean triads (a, b, c) if:

a $n = 7$ ______________________

b $n = 11$ ______________________

c n is any number you choose. What pattern do you notice about b and c?

____

6 A third formula allows n to be odd or even:

$a = 2n + 1, b = 2n^2 + 2n, c = 2n^2 + 2n + 1$

a Substitute $n = 4$ to find (a, b, c).

____

____

b Substitute another value of n.

____

____

c What do you notice about a^2 and $b + c$?

____

d What do you notice about $c - b$?

____

Note: These formulas do not necessarily generate all possible Pythagorean triads.

CHALLENGE: AN ANCIENT CHINESE PUZZLE

A bamboo stick that is 32 cubits high is snapped by the wind and now its tip touches the ground 16 cubits from its base. Where did the bamboo snap?

1 APPLICATIONS OF PYTHAGORAS' THEOREM

NOW IT'S TIME TO USE PYTHAGORAS' THEOREM TO SOLVE PROBLEMS.

WS WORKSHEET

Write all answers correct to 2 decimal places where appropriate.

1 Gurmit calculated the distance across the pond by taking the measurements shown. What is this distance?

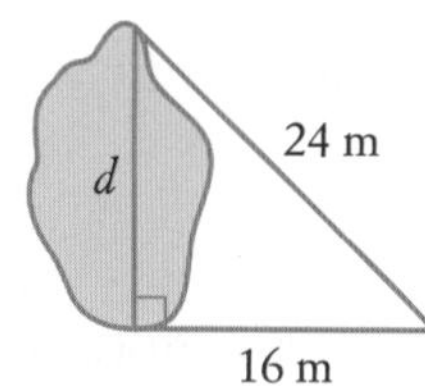

2 Yolanda is moving house and wants to pack a metre-length umbrella into a cube-shaped box of length 60 cm.

Can she pack the umbrella along the length:

a *HG*?

b *HF*?

c *HD*?

d *HC*?

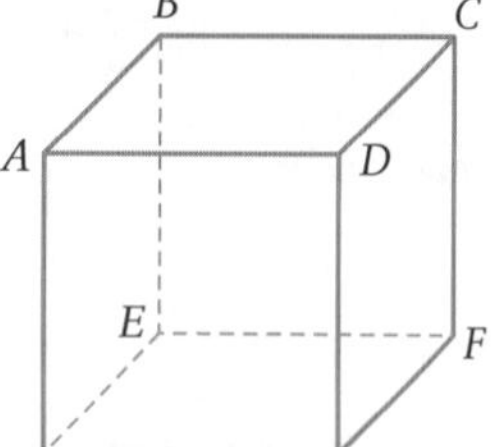

3 TV screen sizes are described by the lengths of their diagonals. Calculate the diagonals of these screens:

Small: 89 cm by 50 cm

Medium: 112 cm by 62 cm

Large: 133 cm by 75 cm

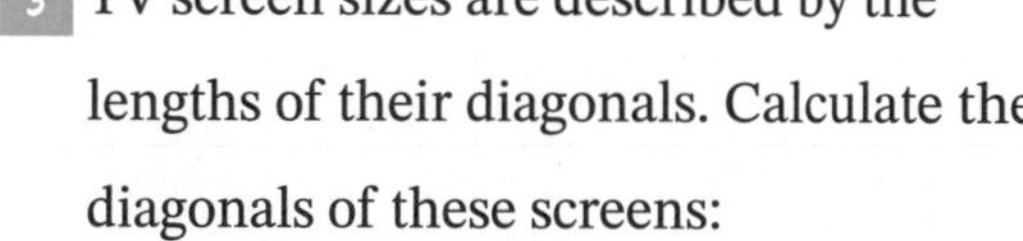

4 A 6 metre ladder leans against a house so that its base is 2 metres out from the bottom of the house. How far up the house does the ladder reach?

5 A yacht leaves Newcastle and sails 160 nautical miles due north. It then turns and sails due east until it is directly 200 nautical miles from Newcastle. How far east did the yacht sail?

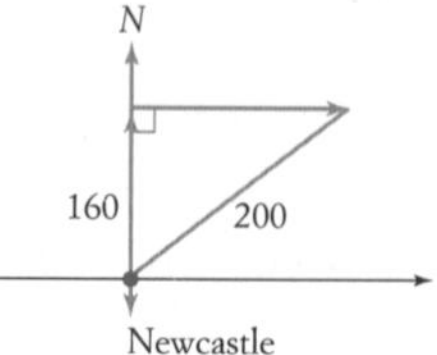

6 Calculate the perimeter of each shape.

a

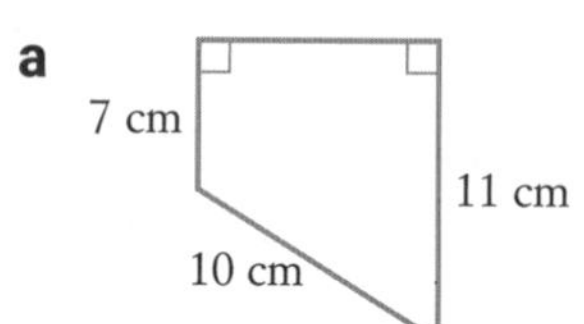

b *ABCD* is a rhombus.

$AC = 12$ cm

$BD = 18$ cm

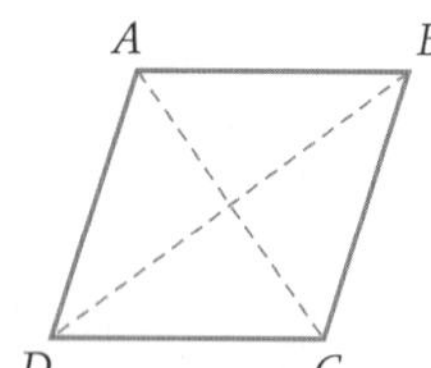

7 Kristine wants to use an old tennis ball can as a pencil case. If the can has a diameter of 7.5 cm and a height of 20 cm, what is the length of the longest pencil which will fit into the can?

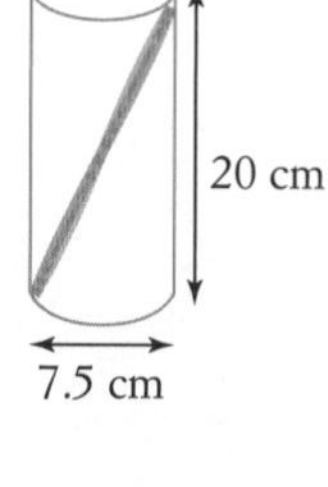

8 If a tent pole is 2 m high and the rope is 2.4 m, how far from the base of the pole should the rope be pegged?

C S F

9780170454520

9 Use a ruler to draw a right-angled triangle with:

a 2 shorter sides 2.5 cm and 6 cm, and measure the length of its hypotenuse;

b a hypotenuse of length $\sqrt{20}$ cm.

10 Ray is staying at Eddie's house when Pete rings to ask them both to the local pool. Ray goes to Pete's house first and they walk to the pool together, while Eddie goes directly to the pool, buying them icy poles at the shops. The map below shows where each of them live.

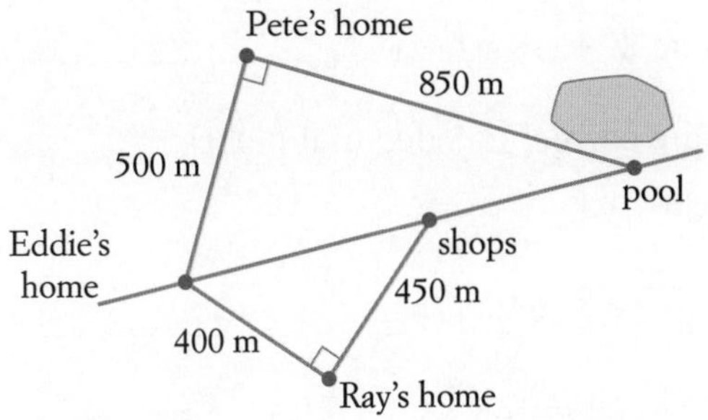

a How much further did Ray walk than Eddie?

b After the swim, Ray walks back to his own house. How far does he walk then?

11 The diagram shows a boy flying a kite. How high is the kite above the ground?

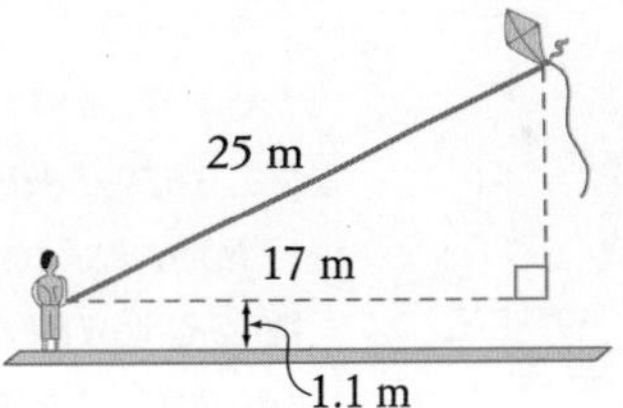

12 If the areas of the square and triangle are equal, then find h and d.

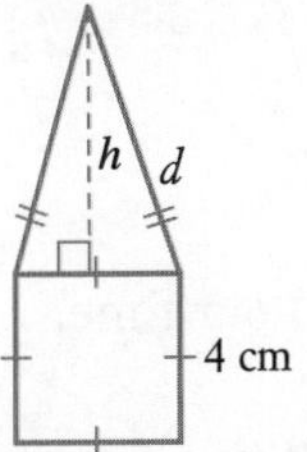

13 **Challenge:** If $ABCD$ is a square with 4 cm sides, find the length of JL.

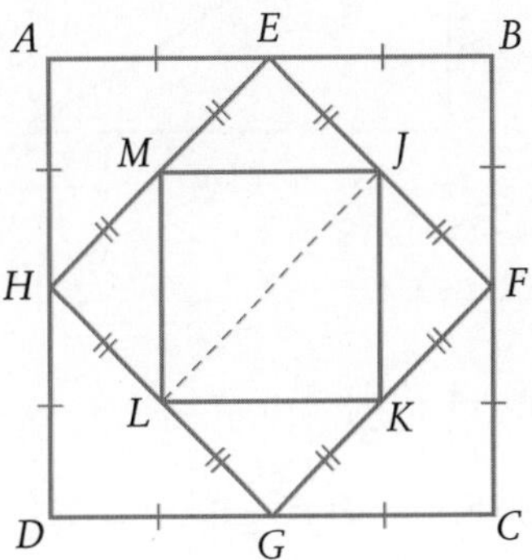

WS WORKSHEET

C S F

1 PYTHAGORAS' THEOREM 1

Name:

Due date:

Parent's signature:

Part A	/ 8 marks
Part B	/ 8 marks
Part C	/ 8 marks
Part D	/ 8 marks
Total	/ 32 marks

THIS IS YOUR WEEKLY HOMEWORK ASSIGNMENT, COVERING THE CURRENT TOPIC AS WELL AS MIXED REVISION. NO CALCULATORS IN PART A!

PART A: MENTAL MATHS

Calculators not allowed

1 Write 21:08 in 12-hour time. ______

2 Complete: 60.8 cm = ______ mm

3 Find the area of this trapezium.

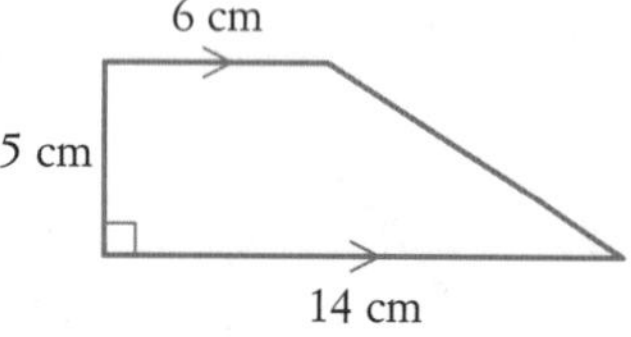

______

4 Evaluate 44×5. ______

5 Factorise $8m - 4m^2$. ______

6 Find the perimeter of this shape.

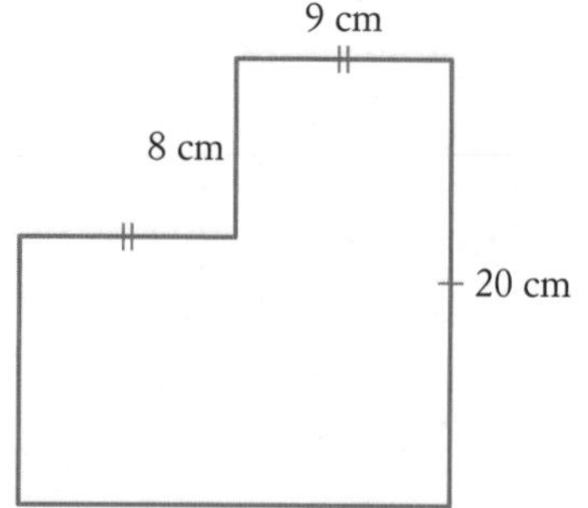

______

7 Evaluate $230 \div 5$. ______

8 Evaluate $\frac{2}{7} - \frac{3}{14}$.

______

PART B: REVIEW

1 Complete: 4120 kg = ______ t

2 Evaluate $8^2 + 3^2$. ______

3 Evaluate correct to 2 decimal places:

a $\sqrt{95}$ ______

b $\sqrt{15^2 - 10^2}$ ______

4 Solve $x - 30 = 145$. ______

5 Find p.

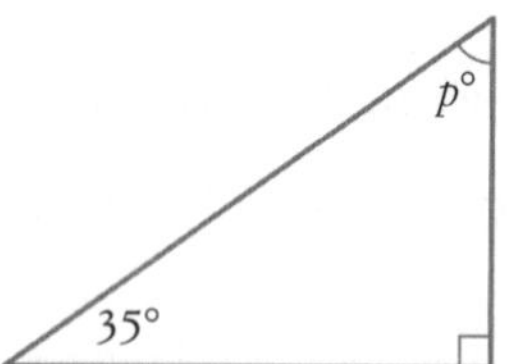

______

6 Solve $2y + 18 = 24$.

______

7 Solve $\frac{m}{12} = 3$. ______

PART C: PRACTICE

› Pythagoras' theorem
› Finding the hypotenuse
› Finding a shorter side

1 **a** Which side of this triangle is the hypotenuse?

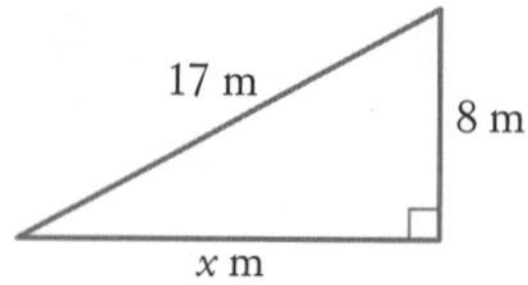

______

C S F

9780170454520

b Find x.

______

2 Find k as a surd.

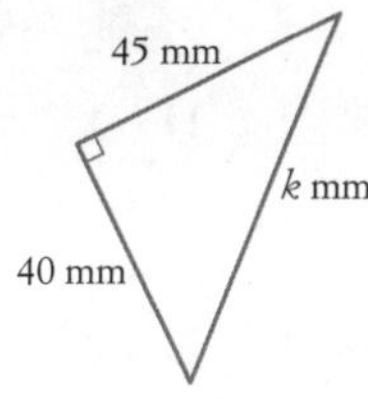

______

3 Find p, correct to one decimal place.

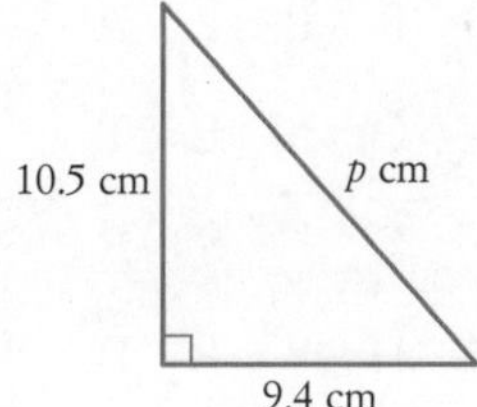

______

4 Select the surds from this list of square roots:

$\sqrt{9}$, $\sqrt{25}$, $\sqrt{49}$, $\sqrt{50}$, $\sqrt{33}$, $\sqrt{69}$.

5 Find y correct to 2 decimal places.

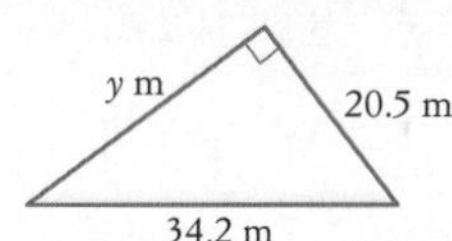

______

6 (2 marks) Construct a right-angled triangle with perpendicular sides of length 25 mm and 60 mm and by measurement find the length of its hypotenuse.

PART D: NUMERACY AND LITERACY

1 Describe the meaning of each word:

a perimeter ______

______

b hypotenuse ______

______

2 Write Pythagoras' theorem for this triangle.

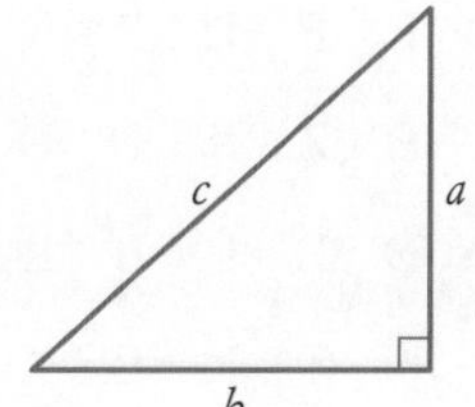

______

3 What name is given to a square root that cannot be expressed as an exact decimal?

______

4 Write 3^2 in words.

______

5 **a** Draw a right-angled isosceles triangle.

b Write the size of each angle in the above triangle.

______

6 Complete: In a right-angled triangle, the square of the hypotenuse is equal to

______

______

1 PYTHAGORAS' THEOREM FIND-A-WORD

FIND THE WORDS FROM THIS TOPIC HIDDEN IN THE PUZZLE.

Y V O B T Q T O J Y Q Z K F O O R P M Y U P I U X

X Q E R A U Q S R U O F G A W C J M E R O E H T A

Q C X T H R E E Z T S E T M T O O R R P N T S U K

Y L P P A U Z A E R A J C I D E D N U O R O N U Y

Y Q P S L A M I C E D L A Y R K D I A G O N A L K

I P E V C P I V C Y A D T C A X E Y E C S X R R R

S L R E V I F U P N K X D X E T A M I X O R P P A

Y K I R K E Z X O D P P Y T H A G O R A S O M E T

Y J M U G D D I B R E V O C S I D F J O E D I S R

P N E Q C N T X E K E M C E Y M H P U O Z H Y B B

T U T J I A R Y N T I G K G C T A J N Y X A B H X

G M E R R E Z J U S L X T G K J I N Z I V J M M G

Y B R R T T M T P K P J L T E U M J O L R I H D T

B E I R D Z I S B C X L C W L E K Y J B X T D A S

U R O U N T C M H T G N E L J Q W R G D C D T I E

I H Q L S E L G N A I R T D E L G N A T H G I R G

S L P B D R Z E S U N E T O P Y H S U R D L W T N

M P U J H V I N A L U M R O F G Y C O Q Z B G F O

U S E T I S O P P O W U E S R E V N O C J W T S L

B V E H Y J N A P Y B J H D E S Y F W Y S G U F C

Find these words in the puzzle above. They appear across, up and down, and diagonal, and can be backwards as well as forwards.

APPLY	APPROXIMATE	AREA	CONVERSE
DECIMAL	DIAGONAL	DISCOVER	EXACT
FIVE	FORMULA	FOUR	HYPOTENUSE
IRRATIONAL	LENGTH	LONGEST	NUMBER
OPPOSITE	PERIMETER	PROOF	PYTHAGORAS
RIGHT-ANGLED	ROOT	ROUNDED	SHORTER
SIDE	SQUARE	SUBSTITUTE	SURD
TEST	THEOREM	THREE	TRIAD
TRIANGLE			

PS PUZZLE SHEET

STARTUP ASSIGNMENT 2 (2)

HI, I'M MS LEE. THIS ASSIGNMENT COVERS BASIC SKILLS AND THINGS YOU SHOULD KNOW BEFORE LEARNING THE NUMBERS TOPIC.

WS WORKSHEET

PART A: BASIC SKILLS

/ 15 marks

1 What is the time 2 hours 20 minutes after 6:45 p.m.?

____

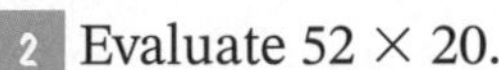

2 Evaluate 52×20.

____

3 What type of angles are these?

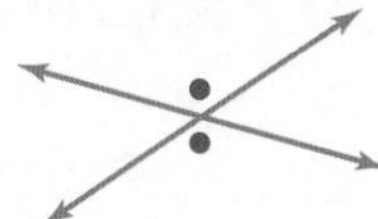

____

4 In which quadrant of the number plane would you find the point $(-3, -5)$? ______________________

5 Evaluate $5.3 + 0.009$. ______________________

6 Find the perimeter of this figure.

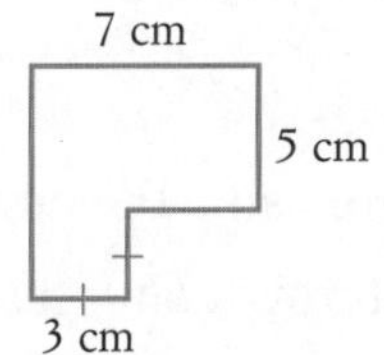

____

7 Complete: 1 kL = ______________________ L

8 Simplify $3x + 4x$. ______________________

9 What is the name of this shape?

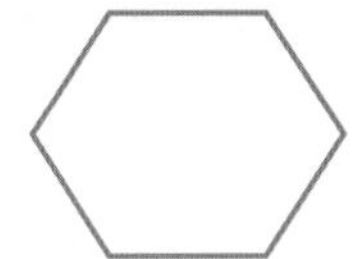

____

10 Complete this pattern:

1, 4, 9, 16, ____, ____

11 If $r = 5$, evaluate $14 - 2r$.

12 Find the area of this triangle.

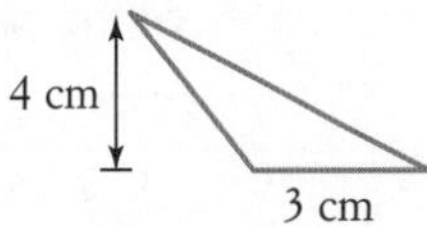

____

13 How many faces has a triangular prism?

____

14 List the factors of 20.

____

15 Draw an obtuse-angled triangle.

PART B: NUMBER

/ 25 marks

16 Convert $\frac{3}{5}$ to a decimal. ______________________

17 Arrange these integers in ascending order:

$-8, 4, 17, -9, 0, -5$

____

18 Find the highest common factor of 8 and 20.

____

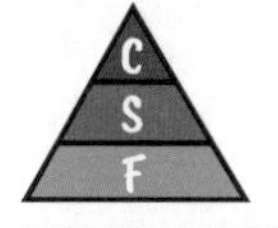

WS WORKSHEET

19 Round 145.236 to the nearest:

a hundredth ______

b whole number ______

20 Evaluate:

a 8^2 ______

b $\sqrt{81}$ ______

c $-6 + 5$ ______

d $-6 - 5$ ______

e 16×20 ______

f $\sqrt{16 \times 4}$ ______

21 Write a prime number between 45 and 50.

______

22 Find the lowest common multiple of 6 and 9.

______

23 Arrange these decimals in descending order:

3.41, 3.14, 3.416, 3.146

______

24 Convert 0.15 to a simplified fraction.

______

25 Evaluate:

a $\sqrt[3]{64}$ ______

b $-7 + (-3)$ ______

c $48 + 16 + 32$ ______

d 15×4 ______

e $(3^2)^2$ ______

f $18 - 8 \div 2$ ______

26 Evaluate 23×11.

______

27 Evaluate $144 \div 6$.

______

28 (2 marks) Draw a factor tree for 175 and express 175 as a product of its prime factors.

PART C: CHALLENGE

Bonus / 3 marks

Make a magic square using the numbers 1 to 9 so that the sum of each row, column and diagonal is 15.

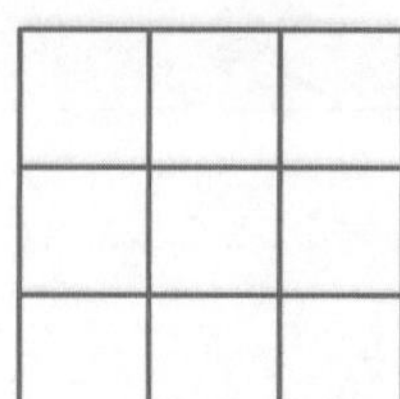

C S F

9780170454520

DECIMAL NUMBER GRIDS 2

DO THESE WITHOUT A CALCULATOR, THEN CHECK YOUR ANSWERS WITH A CALCULATOR!

WS WORKSHEET

Complete each number grid.

1

+	6.8	3.1
5.6		
4.4		

2

+	8.3	2.6
10.7		
2.9		

3

+	4.4	5.6
1.2		
2.6		

4

+	1.12	3.46
2.05		
7.36		

5

+	8.7	1.9	4.8
2.2			
5.8			

6

+	1.15	2.64	2.06
1.07			
7.96			

Top row minus the left hand column in questions **7** and **8**.

7

−	8.4	7.7
2.5		
3.8		

8

−	9.7	6.3
1.4		
4.9		

9

×	1.2	2.4	3.6	3.5
0.2				
0.3				
0.7				

Top row divided by the left hand column in question **10**.

10

÷	8.4	9.6	7.2	3.6
0.2				
0.4				
1.2				
0.6				

② NUMBERS CROSSWORD

DO YOU REMEMBER THE TERMINOLOGY USED IN THIS TOPIC? THIS CROSSWORD WILL HELP, BUT SOME OF THE CLUES ARE TRICKY.

PS PUZZLE SHEET

1 2 3 4 5 6 7 8 9 10 11 12 13 14 15 16 17 18 19 20 21 22 23 24 25 26 27 28 29 30 31

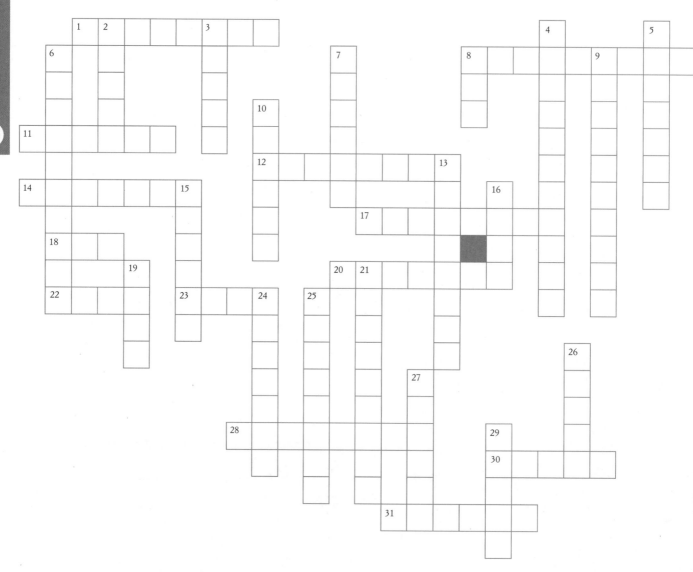

ACROSS

1 ( ) are called ________ symbols

8 From smallest to largest

11 A number that divides into a given number

12 20 is the lowest common ________ of 4 and 10

14 A number whose whole and fraction parts are separated by a dot

17 Another name for highest common factor is g ________ common divisor

18 The value of a number raised to the power of 0

20 The answer to a multiplication

22 To find a number's prime factors, use a factor ________

23 A root such as $\sqrt{7}$ whose exact decimal value cannot be found

28 5^4 is an example of index ________

30 Doing × before + is an example of ________ of operations

31 1.708 has 3 decimal ________

DOWN

2 To multiply a decimal by 100, move the decimal point 2 places to the ________

3 Another name for power beginning with i

4 A decimal that is not recurring is ________

5 Positive or negative whole number

6 $\sqrt{\ }$ (2 words)

7 When adding or subtracting decimals on paper, keep decimal ________ below one another

8 To find a sum, you ________

9 The answer to a subtraction

10 The C in LCM

13 E-word means to find the value

15 The L in LCM

16 To divide a decimal by 10, move the decimal point one place to the ________

19 An integer that is neither positive nor negative

21 A repeating decimal

24 Another name for factor, the D in GCD

25 To simplify $2^9 \div 2^3$, ________ the powers

26 A number that has only 2 factors

27 ________ calculation means operating with numbers using your mind

29 The 4 in 10^4

2 INTEGERS

JUST LIKE IN SPORT, DRILL AND PRACTICE ARE IMPORTANT IN MATHS. PART B OF THIS HOMEWORK ASSIGNMENT IS REVIEW OF PREVIOUS WORK, PART C IS PRACTICE OF THIS WEEK'S WORK.

Name:

Due date:

Parent's signature:

Part A	/ 8 marks
Part B	/ 8 marks
Part C	/ 8 marks
Part D	/ 8 marks
Total	/ 32 marks

PART A: MENTAL MATHS

Calculators not allowed

1 Evaluate each expression.

a $\sqrt{64}$ ________________

b 48×11 ________________

2 Draw an isosceles triangle.

3 Complete: $\frac{2}{3} = \frac{}{18}$

4 Solve the equation $4x + 1 = 25$

____

5 A rectangle is 8 cm long and 3 cm wide. Find:

a its perimeter

____

b its area

____

6 How many degrees in a revolution?

____

PART B: REVIEW

1 Complete, using a $<$ or $>$ sign:

a 4 ______ -2

b -3 ______ 5

c -2 ______ -1

d -3 ______ -6

2 (2 marks) Write these integers in descending order:

$7 \quad -5 \quad -2 \quad 0 \quad -9 \quad -4 \quad 8 \quad -7$

____

3 What integer is:

a 8 more than 3? ________________

b 8 less than 3? ________________

C S F

9780170454520

PART C: PRACTICE

› Adding and subtracting integers
› Multiplying and dividing integers

1 Evaluate each expression.

a $6 + (-6)$ ______

b $-2 - 9$ ______

c $-4 - (-5)$ ______

d $-8 + 3 + (-5)$ ______

e $-1 + (-1) - (-10)$ ______

f $-7 \times (-2)$ ______

g $16 \div (-8)$ ______

h $5 + (-4) \times [12 \div (-6)]$ ______

PART D: NUMERACY AND LITERACY

1 Circle all the numbers below that are **not** integers.

$18, -8, 6.5, 3, 0, -4, \frac{1}{3}$

2 The temperature decreased from 7°C last night to −2°C this morning. By how many degrees did it decrease?

______

3 What is the sign of the quotient of 2 negative integers?

______

4 At the snow, it was −2°C at 6 a.m.

The temperature increased by 6°C by 11 a.m., then by another 2°C by 1 p.m. It then decreased by 3°C at 5 p.m. and another 3°C by 9 p.m.

Find the temperature at:

a 1 p.m.

______

b 5 p.m.

______

c 9 p.m.

______

5 Write 3 integers that have a product of −20.

______

6 Is a positive integer minus a negative integer always positive, always negative or could be either positive or negative?

______

HW HOMEWORK

2 DECIMALS 1

Name:

Due date:

Parent's signature:

Part A	/ 8 marks
Part B	/ 8 marks
Part C	/ 8 marks
Part D	/ 8 marks
Total	/ 32 marks

PART D QUESTIONS CAN BE COMPLEX BECAUSE THEY ASK YOU TO WRITE ABOUT YOUR MATHS, USING THE RIGHT TERMINOLOGY.

HW HOMEWORK

PART A: MENTAL MATHS

Calculators not allowed

1 Evaluate each expression.

a $\sqrt[3]{27}$ ____________

b $\frac{5}{8} \div \frac{15}{8}$ ____________

2 What is the probability that a person chosen at random has a birthday in a month later than April?

____

3 If $n = 2$, then evaluate $10 + 6n$.

____

4 Write the coordinates of a point on the x-axis of the number plane.

____

5 Convert $\frac{17}{40}$ to a percentage.

____

6 Complete: 1 hour = ________ seconds.

7 Mark a pair of vertically opposite angles.

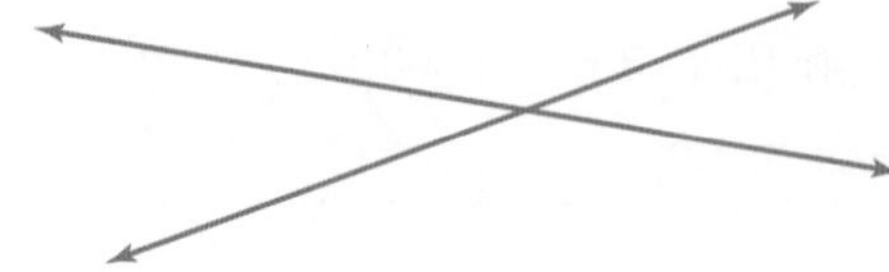

PART B: REVIEW

1 Round 1937 to the nearest ten.

____

2 Convert 0.28 to a simple fraction.

____

3 How many decimal places has 3.1416?

____

4 Evaluate each expression.

a $8 + 3 \times 4$ ____________

b $50 - [-2 \times 6 \div (-3)]$ ____________

c $3600 \div 100$ ____________

5 (2 marks) Write these decimals in ascending order.

0.8 0.2 1.5 0.03 2.04 0.15

____

C S F

9780170454520

PART C: PRACTICE

› Rounding decimals
› Adding, subtracting and multiplying decimals

1 Round:

a 3.2851 correct to 2 decimal places

b 8.92374 correct to 3 decimal places

2 Evaluate each expression.

a $52.67 + 124.082$

b $14.651 - 3.98$

c $386.94 - 28.926$

d 24.6×7

e $(2.3)^2$

f 564.28×1.5

PART D: NUMERACY AND LITERACY

1 Write a decimal that can be rounded to 16.54.

2 (2 marks) Complete.

a Rounding to the nearest tenth means rounding to ______ decimal place(s)

b When adding and subtracting decimals, keep the decimal ______ underneath ______ other.

3 (2 marks) Calculate the sum of 4.6 and 0.56 divided by the difference between 6.2 and 4.2.

4 (3 marks) Lily bought the following items:

3 packets of biscuits @ \$2.80/packet

5 kg of apples @ \$3.20/kg

4 L of ice cream @ \$1.25/Litre

2 kg of sugar @ \$1.80/kg

a What will the groceries cost her?

b How much change will she receive from a \$50 note?

2 DECIMALS 2

IF YOU CAN SOLVE THESE QUESTIONS SUCCESSFULLY, THEN I THINK YOU'VE MASTERED DECIMALS.

Name:

Due date:

Parent's signature:

Part A	/ 8 marks
Part B	/ 8 marks
Part C	/ 8 marks
Part D	/ 8 marks
Total	/ 32 marks

PART A: MENTAL MATHS

Calculators not allowed

1 Complete: $\frac{4}{7} = \frac{}{28}$.

2 Find the range of this set of data.

12, 5, 9, 7, 8, 7, 10.

3 Simplify $4 \times r \times 9 \times v$.

4 Write an algebraic expression for how many times 7 divides into k.

5 Find the value of d if the perimeter of this rectangle is 46 m.

d m

9 m

6 Complete: 12 km = ______ m

7 Convert $\frac{12}{5}$ to a mixed numeral.

8 Draw a hexagon.

PART B: REVIEW

1 Round 4.5971 correct to:

a the nearest tenth

b 3 decimal places

2 Write down a decimal between 7.04 and 7.048.

3 Convert 0.85 to a simple fraction.

4 Evaluate each expression.

a $2.7 + 9.03$

b 0.5×0.05

c $11.6 - 4.756$

d $2890 \div 5$

C S F

HW HOMEWORK

9780170454520

PART C: PRACTICE

› Dividing decimals
› Terminating and recurring decimals
› Decimals revision

1 Evaluate $36.246 \div 6$.

2 **a** Complete: $8.412 \div 0.04 =$ ________ $\div 4$.

b Evaluate $8.412 \div 0.04$.

3 Rewrite the recurring decimal

15.137137137 ...

using dot notation.

4 Convert each fraction to a decimal.

a $\frac{7}{8}$

b $\frac{5}{6}$

5 Evaluate each expression.

a $2.75 \div 1000$

b $6.8 \times 0.4 \div 0.02$

PART D: NUMERACY AND LITERACY

1 What is a **recurring decimal**?

Give an example of one.

2 Write a decimal that could be rounded down to 2.25.

3 (2 marks) Explain how $37.458 \div 0.9$ can be evaluated by moving decimal places first and then find the answer.

4 **a** The cost of 5 kg of honey is \$19.75.

What is the cost of 1 kg of honey?

b How many kg of honey can be bought for \$16.59?

5 Given that $35 \times 14 = 490$, evaluate without using a calculator:

a 3.5×1.4

b 0.35×14

HW HOMEWORK

2 MENTAL CALCULATION AND POWERS

THIS ASSIGNMENT COVERS CALCULATING WITHOUT CALCULATORS, AND USING POWERS AND ROOTS.

Name:

Due date:

Parent's signature:

Part A	/ 8 marks
Part B	/ 8 marks
Part C	/ 8 marks
Part D	/ 8 marks
Total	/ 32 marks

HW HOMEWORK

PART A: MENTAL MATHS

Calculators not allowed

1 How many days in August? ____

2 Simplify $\frac{12}{20}$. ____

3 Evaluate $8 + (-6)$. ____

4 Find the median of these data values:

12, 5, 9, 7, 8, 7, 10. ____

5 Describe an acute angle in words.

6 Find the area of this triangle.

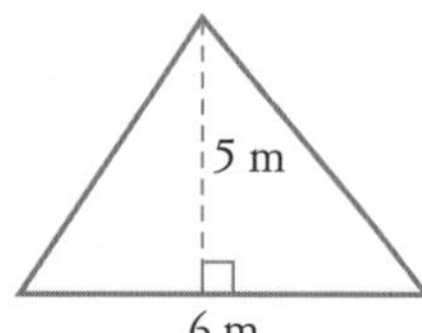

7 Convert 0.35 to a simplified fraction.

8 Write an algebraic expression for the product of b and c.

PART B: REVIEW

1 Find the square number between 40 and 50.

2 Find the difference between 56 and 28.

3 Write each expression using index notation.

a $6 \times 6 \times 6 \times 6 \times 6 \times 6$ ____

b $4 \times 4 \times 4 \times 4 \times 4 \times 4 \times 4 \times 4$ ____

4 Evaluate each expression.

a 5^4 ____

b $3^2 \times 4^3$ ____

c $\sqrt{729}$ ____

d $\sqrt{9 \times 25}$ ____

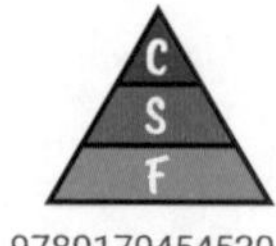

9780170454520

PART C: PRACTICE

› Mental calculation
› Powers and roots
› Multiplying and dividing terms with the same base

1 Evaluate each expression without using a calculator.

a $829 - 38$

______

______

b 48×50

______

______

c $6885 \div 9$

______

______

2 Evaluate each expression.

a $(-2)^6$ ______

b $\sqrt[3]{343}$ ______

3 Use index notation to simplify each expression.

a $2^5 \times 2^6$ ______

b $3^8 \div 3^3$ ______

c $8^6 \div 8^4 \times 8^2$ ______

PART D: NUMERACY AND LITERACY

1 **a** Write 4^5 in words.

______

______

b In 4^5, what is the index? ______

2 Alice bought 3 items at the supermarket:

- Milk for \$2.65
- A loaf of bread for \$3.28
- Chocolates for \$8.75

a Calculate, correct to the nearest 5 cents, the total cost of these items.

______

______

b If she pays with \$20 cash, how much change will she receive?

______

______

3 Complete: When dividing terms with the same base, we ______ the powers.

4 (2 marks) Explain how to calculate 14×12 mentally.

______

______

______

______

5 The numbers of students in 6 Year 8 classes are 29, 26, 30, 28, 26 and 29. If the classes were reorganised so that every class was the same size, how many students would be in each class?

______

______

______

______

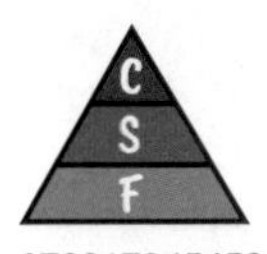

2 POWERS AND DIVISIBILITY

MORE WORK ON POWERS AND ROOTS HERE. KEEP PRACTISING, YOU'LL GET THERE!

Name:

Due date:

Parent's signature:

Part A	/ 8 marks
Part B	/ 8 marks
Part C	/ 8 marks
Part D	/ 8 marks
Total	/ 32 marks

HW HOMEWORK

PART A: MENTAL MATHS

Calculators not allowed

1 Complete: 4.8 m = ____________ mm.

2 Simplify $\frac{25}{15}$. ____________

3 Write 8% as a decimal. ____________

4 Write an algebraic expression for the sum of a, b and 7. ____________

5 Shade in 0.9 of this shape:

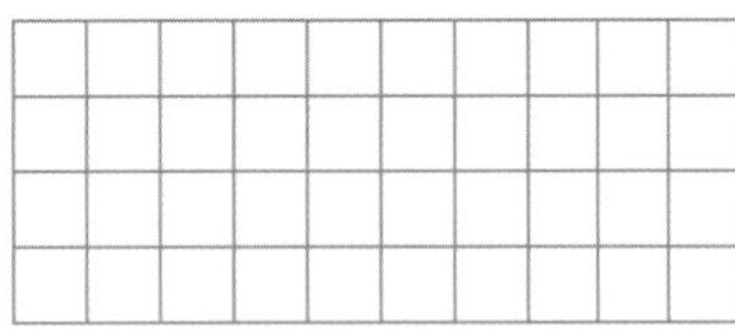

6 Evaluate 82 − 19. ____________

7 What type of angle is this?

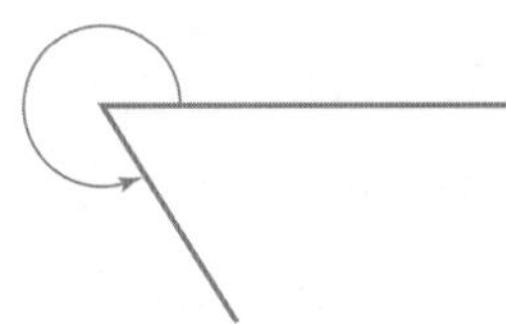

____

8 Find the quotient of 65 and 5.

____

PART B: REVIEW

1 Evaluate each expression.

a 3^4 ____________

b $(-7)^2$ ____________

c $\sqrt{64}$ ____________

d $\sqrt[3]{64}$ ____________

2 Use index notation to simplify each expression.

a $5^3 \times 5^7$ ____________

b $\frac{9^8}{9^4}$ ____________

3 (2 marks) List the factors of 20.

____

____

C S F

9780170454520

PART C: PRACTICE

› Power of a power, the zero index
› Divisibility tests

1 Use index notation to simplify each expression.

a $(3^2)^3$ ____________

b $(5^3)^4$ ____________

2 Evaluate each expression.

a 10^0 ____________

b $(5 + 4)^0$ ____________

c $2 - 7^0$ ____________

3 Test whether 684 is divisible by:

a 3

____

____

b 5

____

____

c 6

____

____

PART D: NUMERACY AND LITERACY

1 **a** Evaluate $\sqrt[3]{15\,625}$. ____________

b Explain what the cube root of 15 625 means.

____

____

____

2 Complete: If a number is divisible by 6, then it must be divisible by ________ and be divisible by ________.

3 **a** Write a 5-digit number that is divisible by 9.

____

b Describe the divisibility test for 9, using the number you gave in part **a** as an example.

____

____

____

____

4 **a** Use index notation to simplify $\frac{4^5}{4^5}$ ____________

b What is the answer if you divide any answer by itself? ____________

c What does this show about any number raised to the power of 0?

____

____

HW HOMEWORK

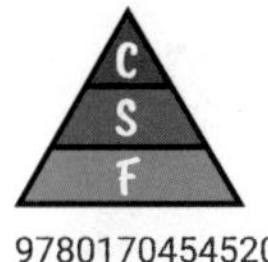

2 PRIME FACTORS

Name:

Due date:

Parent's signature:

Part A	/ 8 marks
Part B	/ 8 marks
Part C	/ 8 marks
Part D	/ 8 marks
Total	/ 32 marks

HAVE YOU WORKED OUT WHY SOME QUESTIONS ARE BLUE AND SOME ARE GREEN? THERE'S EVEN SOME RED ONES NEXT PAGE. YOU'LL NEED MORE TIME TO DO THOSE.

PART A: MENTAL MATHS

Calculators not allowed

1 Evaluate each expression:

a 5^3 ____

b $\frac{4}{9}+\frac{2}{9}$ ____

c 58×9 ____

d the sum of 2.5 and 8.74

____

2 What is another name for a 90° angle?

____

3 If $a = 8$, then evaluate $10 - 2a$.

____

4 Find the mode of these data values:

12, 5, 9, 7, 8, 7, 10.

____

5 If a coin is tossed, what is the percentage probability that it comes up tails?

____

PART B: REVIEW

1 List all of the factors of:

a 18

____

b 11

____

2 List the first 6 multiples of 4.

____

3 (2 marks) Circle the 3 prime numbers from this list:

23 54 35 47 61 75 87

4 Find the highest common factor of 20 and 12.

____

5 Find the lowest common multiple of 20 and 12.

____

6 List all of the composite numbers between 10 and 20.

____

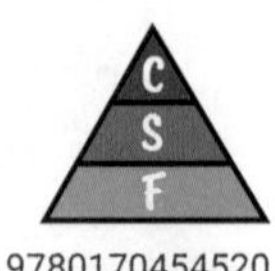

HOMEWORK

9780170454520

PART C: PRACTICE

› Prime factors
› Highest common factor
› Lowest common multiple

1 (6 marks) Draw a factor tree for each number and write the number as a product of its prime factors.

a 72

b 120

c 84

2 Use your answers from question **1** to find:

a the highest common factor for 72 and 120

b the lowest common multiple of 84 and 72

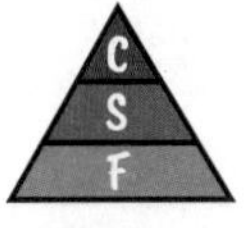

PART D: NUMERACY AND LITERACY

1 **a** Explain what a prime number is, giving an example.

b How many factors has any prime number?

c What number is neither prime nor composite?

2 What does HCF stand for?

3 When written as a product of their prime factors, $64 = 2^6$ and $96 = 2^5 \times 3$.

Use this to find:

a the highest common factor of 64 and 96

b the lowest common multiple of 64 and 96

4 **a** Most numbers have an even number of factors. Find a number that has an **odd** number of factors.

b What type of numbers have an odd number of factors?

HW HOMEWORK

2 POWERS AND PRIME FACTORS

THIS IS THE LAST ASSIGNMENT FOR THIS TOPIC. THE QUESTIONS AREN'T EASY, SO ASK FOR HELP IF YOU GET STUCK.

Name:

Due date:

Parent's signature:

Part	Marks
Part A	/ 8 marks
Part B	/ 8 marks
Part C	/ 8 marks
Part D	/ 8 marks
Total	/ 32 marks

PART A: MENTAL MATHS

Calculators not allowed

1 What is 11 squared? ____________

2 Evaluate 12×13. ____________

3 What is a straight angle?

____

____

4 Find, correct to one decimal place, the mean of these numbers:

12, 5, 9, 7, 8, 7, 10.

____

5 Shade 40% of this shape.

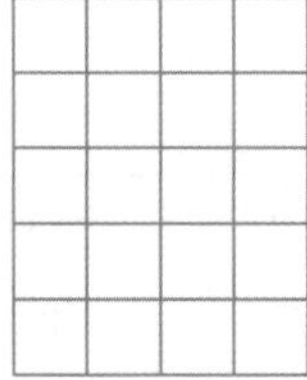

6 Convert 16:40 (in 24-hour time) to 12-hour time. ____________

7 Convert 6% to a simple fraction.

____

8 Write an algebraic expression for the number that is 10 less than double y.

____

PART B: REVIEW

1 Use index notation to simplify each expression.

a $8^4 \times 8^5$ ____________

b $4^8 \div 4^4$ ____________

c $(3^5)^3$ ____________

d $6^0 + 3^0$ ____________

2 (2 marks) Write down all the prime numbers between 1 and 20.

____

____

3 (2 marks) List the factors of 16 and 32, and find the highest common factor of 16 and 32.

____

____

____

____

C S F

9780170454520

PART C: REVIEW

1 Evaluate $\sqrt[3]{512} - \sqrt{169}$. ____________

2 (6 marks) Draw a factor tree for each number and write the number as a product of its prime factors.

a 60

b 140

3 Use your answers from above to find the lowest common multiple of 60 and 140.

____

PART D: NUMERACY AND LITERACY

1 Explain how you used the factor trees to find the lowest common multiple of 60 and 140 in question **3** of Part C.

____

____

____

2 (2 marks) Write '10 to the power of 4' in index notation, and find its value.

____

3 (2 marks) Complete: The ______ ______ of 125 is ______ because 5^3 = ______.

4 Explain what a composite number is, giving an example.

____

____

____

5 Complete: When raising a term with a power to another power, such as $(6^3)^2$, we ______ the powers.

6 Find the largest number that is a factor of both 28 and 42.

____

____

____

HW HOMEWORK

C S F

3 STARTUP ASSIGNMENT 3

THIS ASSIGNMENT CAN BE DONE AT THE START OF A TOPIC (ALGEBRA), BECAUSE IT REVISES SKILLS THAT YOU'LL NEED TO LEARN THE TOPIC.

WS WORKSHEET

PART A: BASIC SKILLS

/ 15 marks

1 Complete: 4.2 t = ______ kg.

2 Find a prime number between 21 and 30.

______

3 Draw a hexagon.

4 What is the complementary angle to 70°?

______

5 Evaluate 15 ÷ 17 correct to 3 decimal places.

______

6 **a** Name this solid shape.

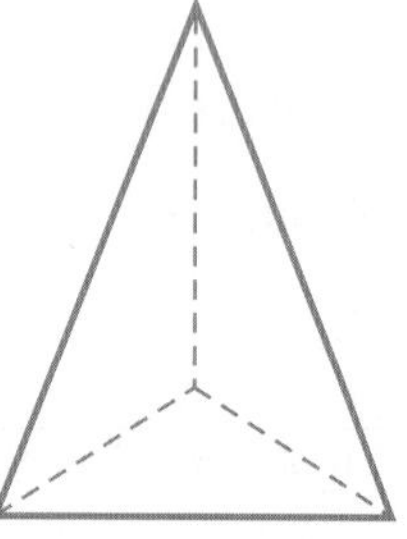

______

b How many faces has this solid? ______

7 What is the point (0, 0) on a number plane called? ______

8 Calculate $\sqrt[3]{343}$. ______

9 Find the square number between 30 and 40.

______

10 What is the size of one unit on this scale?

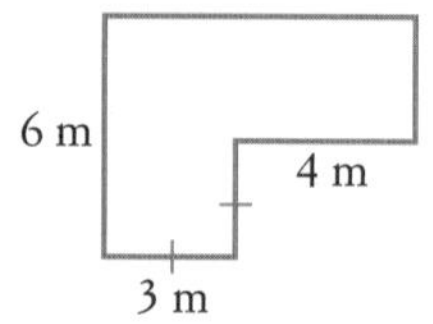

______

11 Convert $\frac{21}{28}$ to a percentage. ______

12 True or false: $20 \div 4 > \sqrt{25}$? ______

13 For this figure, find its:

a perimeter ______

b area ______

______

PART B: NUMBER AND ALGEBRA

/ 25 marks

14 Evaluate each expression.

a $8 \times (-3) - 4$ ______

b $-16 \div (-2)$ ______

c $-3 + 5 \times 2$ ______

d $(-6)^2$ ______

e $2 \times (9 - 4)$ ______

f $2 \times 9 + 2 \times (-4)$ ______

15 List the factors of 16.

______

C S F

9780170454520

16 What is the product of 6 and 9?

17 How many minutes in 4 hours?

18 Find the highest common factor of 18 and 30.

19 If $d = 7$, then evaluate:

a $4d + 5$

b $20 - 3d$

20 Simplify each algebraic expression.

a $7h - 5h$

b $r \times r \times r \times r$

c $6 \times x - 4$

d $6 \times x \times 4$

21 Write an algebraic expression for:

a the product of 5 and y

b the next consecutive number after n.

22 What is the difference between 27 and 9?

23 True or false?

a $a \times b = b \times a$

b $a + b = b + a$

24 If $y = 4x + 6$, find y when $x = 8$.

25 Find the lowest common multiple of 9 and 12.

26 If $p = 16$ and $q = 5$, evaluate $2p - q^2$.

27 Draw a factor tree for 60, then use it to write 60 as a product of its prime factors.

PART C: CHALLENGE

Bonus / 3 marks

If you divide Ben's age by 2, 3, 4, 5 or 6, the remainder is 1.

If Ben is less than 100 years old, how old is he?

C S F

3 WHAT'S THE EXPRESSION?

ALGEBRA IS ABOUT CONVERTING A PROBLEM INTO MATHEMATICAL SYMBOLS, SO CHANGE THESE WORDS INTO ALGEBRA.

Match each expression given in words with its correct algebraic expression shown below. Then use the matched question numbers and capital letters to decode the answer to the riddle on the next page.

1 The next consecutive number after x.

2 The change (in dollars) from \$20 after spending \$$y$.

3 Triple y.

4 The product of x and y.

5 Half of y.

6 The square root of x.

7 The sum of x and y.

8 12 less than y.

9 y times y times y.

10 Two times x.

11 Increase x by 4.

12 The average of x and y.

13 x reduced by 4.

14 The year 20 years from now if this year is y.

15 The difference between 12 and y, where 12 is larger.

16 The quotient of y and 12.

17 Double x plus y.

18 The even number before y if y is even.

19 Two divided by x.

20 x increased by 2.

21 The next odd number after y if y is even.

22 x divided by 2.

23 The change (in dollars) from \$$y$ after buying 2 hamburgers at \$$x$ each.

24 Decrease 4 by x.

25 Darren's age 3 years ago if he is y years now.

26 The number 10 more than y.

27 Subtract x from y.

28 The cost of 5 bananas (in dollars) if one costs x.

29 x multiplied by itself.

30 One share (in dollars) when \$$x$ is divided evenly among 5 people.

A	$5x$	**V**	$x+y$	**P**	$x-4$	**T**	$x+4$
C	$\frac{y}{2}$	**M**	$y+1$	**D**	$y-3$	**Y**	xy
J	y^3	**L**	$y-x$	**Q**	$20-y$	**N**	$y+20$
W	$y-12$	**N**	$4-x$	**H**	$x+2$	**Z**	$x+1$
K	$y+10$	**E**	$3y$	**R**	$\frac{x}{5}$	**P**	$y-2$
G	x^2	**O**	$2x$	**S**	$\sqrt{x}$	**I**	$\frac{y}{12}$
X	$\frac{2}{x}$	**O**	$\frac{x}{2}$	**F**	$y-2x$	**U**	$\frac{x+y}{2}$
		T	$12-y$	**E**	$2x+y$		

When Mr Point asked Jake, 'If Mum gives you x \$1 coins and Dad gives you y \$2 coins, what is an expression for the total amount of dollars that you'd get?', what did Jake say?

16 , 27 27 | 29 16 7 17 | 4 10 12 | 11 8 10

17 19 18 30 3 6 6 16 10 14 6 : | 10 14 3 | 16 6

' 16 ' | 25 | 22 17 | 20 28 13 18 4 | 28 6 | 27 28 30 30 4 ' .

10 20 3 | 10 11 20 3 30 | 16 6 | ' 16 ' | 7 17 | 29 10 11

22 12 5 26 27 17 4 , | 6 | 5 20 28 24 5 3 , !

WS WORKSHEET

3 COLLECTING LIKE TERMS

1 $x + 5x$

2 $3y + 8y$

3 $6a - a$

4 $9t - 4t$

5 $3ab + 6ab$

6 $8d + 12d$

7 $7st + 3st$

8 $2x + x$

9 $5a^2 + 4a^2$

10 $3t - 2t$

11 $4ab + 4ab$

12 $5ty - 3ty$

13 $2mp - mp$

14 $x + 4x - 3x$

15 $w + w + 5w$

16 $2c - 5c$

17 $3r - 3r$

18 $2m + 3m - m$

19 $e - 2e$

20 $14h + 6 - 12h$

21 $y + y + y$

22 $4c + d + 4c - 3d$

23 $7p + 2 + 3p + 7$

24 $5s - s + 2s$

25 $4y^2 + y^2$

26 $3d + d - d$

27 $a + a + 2f$

28 $c + 12d + 2c - 5d$

29 $6 + 3h + 4h + 1$

30 $6u^2 - 3u^2 + 2u^2$

31 $4a + 2a - 6b - 4b$

32 $5g - 4 + 2g + 3$

33 $10x + 6b - 5b - 7x$

34 $12m + 2n + 8m - 7n$

35 $-4a + 8a^2 + 5a + 3a^2$

36 $y + z + y - z$

37 $12w - 6w + 4wt + 3wt$

38 $4a + 6a + 2b - 3a$

39 $x^2 + 3y + y + x^2$

40 $6f + 3 + 2f + 1$

41 $5y^2 + 4x - 2y^2 + 3x$

42 $4k + p - 3k + p$

43 $x + y - x$

44 $8y - 2 - 6y + 1$

45 $2y + 3m - y - m$

46 $-8 - 2q + 5q + 9$

47 $5r^2 + 3r^2 - r - 2r$

48 $3a - b + 3a$

49 $5t + 3t - 3v - 2v$

50 $18 - k + 4k - 10$

51 $14b - 2 + 3b - 6$

52 $4y + 2z - 3y + z$

53 $2x + 5 - 3x - 2$

54 $2ab + ab + 3b$

55 $2p^2 + 3p - p^2 + p$

56 $3u^2 + 5 - u^2 - 4$

57 $-5a + 2 + 3a + 3$

58 $8x - x - 8$

59 $10 + 6x + 3x - 7$

60 $9p - 3q - 7p - 2q$

61 $8t + 4w - 8t + 2w$

62 $15x + 3x - 5 + 5$

63 $-4p + 6p + a - 4a$

64 $6c + 7d - d - 5c$

62 $9jk - 5k - 2kj + 2k$

66 $5b - 5 - 12b + 5$

67 $-3x + 7y - 7y + 6x$

68 $-5y^2 + yz + 3yz - 2y^2$

69 $2y + 5 - 6y + 2$

70 $w^2 - 6u + 2w^2 + 9u$

71 $2bc + 4cd - bc + 3cd$

72 $-8ef + 2f^2 + 5f^2 + 10ef$

C S F

9780170454520

HI, ZINA HERE. JOIN UP QUESTIONS AND ANSWERS USING A RULER TO SOLVE THIS PUZZLE.

PS PUZZLE SHEET

In the year 825 CE, an Arabic mathematician published a book in Baghdad, titled *Kitab al-jabr wa al-muqabalah* or *The Science of Restoration and Reduction*. From 'al-jabr', we get the word 'algebra'. This book had a great influence on European mathematics of that time. So who was this mathematician?

Factorise the expressions in the left column. Draw a line from each expression on the left to its answer in the right column. When you have finished, start from the top and write all 11 letters that have **not** been crossed. This will give you the Arabic mathematician's name.

$2a + 6$ •	A G	• $-5(a + 1)$
$3a + 18$ •	B	• $a(a + b)$
$5a - 15$ •	L P A L	• $2(a + 3)$
$6a + 14$ •		• $a(5 - a)$
$-5a - 5$ •	K O H	• $5(a - 3)$
$a^2b - ab$ •	F W	• $3a(2b + c)$
$12 - 4a$ •	C	• $3(a + 6)$
$15 - 10a$ •	Z	• $3a(4 + b - 2c)$
$5a - a^2$ •	D E A A	• $-4(3a - 5)$
$a^2b - ab^2$ •	F R	• $ab(a - b)$
$5ay - 2a$ •	I	• $a^2(1 - a)$
$-12a + 20$ •	N	• $ab(a - 1)$
$-5a + 15$ •	M E	• $-a^2(3 + a)$
$a^2 + a$ •	J	• $-5(a - 3)$
$a^2 + ab$ •	Z U	• $5(3 - 2a)$
$6ab + 3ac$ •	M	• $a(5y - 2)$
$a^2 - a^3$ •	S	• $a(a + 1)$
$14 - 21ab$ •	Q	• $7(2 - 3ab)$
$-3a^2 - a^3$ •	C	• $2(3a + 7)$
$12a + 3ab - 6ac$ •	I	• $4(3 - a)$

___ ___ - ___ ___ ___ ___ ___ ___ ___ ___ ___

3 ALGEBRAIC NOTATION AND SUBSTITUTION

HEY, I'M MITCH. THIS ASSIGNMENT COVERS MENTAL MATHS, ALGEBRA AND SUBSTITUTION.

Name:

Due date:

Parent's signature:

Part A	/ 8 marks
Part B	/ 8 marks
Part C	/ 8 marks
Part D	/ 8 marks
Total	/ 32 marks

PART A: MENTAL MATHS

Calculators not allowed

1 Evaluate each expression.

a 8^3 ______

b $12.6 - 0.4 \times 0.7$ ______

2 Draw a triangular prism.

3 Write 0.38 as a percentage. ______

4 Solve the equation $4n - 12 = 16$.

______

______

5 Find the area of this triangle.

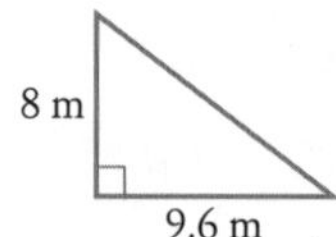

6 Complete: 4.2 kg = ______ g

7 (2 marks) Find a and b.

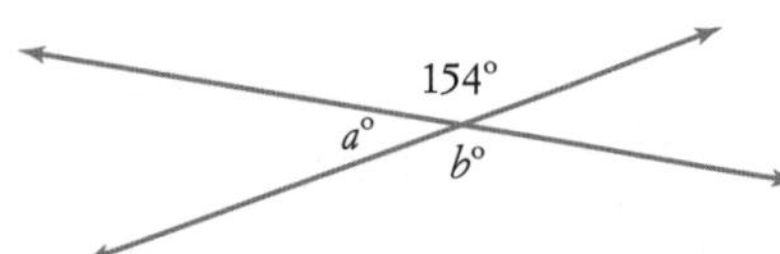

PART B: REVIEW

1 Evaluate each expression.

a $2 \times 8 - 10$

______

b $5 \times (-3) + 4$

______

2 Find the product of 9 and 8.

______

3 Simplify each expression.

a $b + b + b$

______

b $b \times b \times 2$

______

c $b - b$

______

4 If $m = -6$, evaluate each expression.

a $4m + 6$

______

b $10 - m$

______

C S F

HW HOMEWORK

9780170454520

PART C: PRACTICE

› Variables
› From words to algebraic expressions
› Substitution

1 Simplify each expression.

a $9 \times y \times y \times 4$

b $2n - n$

c $16 + r \div 2$

2 Write an algebraic expression for:

a the difference between 10 and a

b 4 more than double b

c the square root of half of c

3 If $m = 3$ and $p = -5$, find the value of:

a $m^2 + p^2$

b $(m + p)^2$

C
S
F

PART D: NUMERACY AND LITERACY

1 Write an algebraic expression for:

a the cost of one ice cream if 4 cost \$$p$

b Andrea's age k years ago if she is 14 this year

2 a Write in words the meaning of the expression $\frac{5d}{3}$.

b What is the value of $\frac{5d}{3}$ if $d = -12$?

3 Explain why:

a $1x = x$

b $0x = 0$

4 The angle sum, A degrees, of a polygon with n sides is given by the formula $A = 180n - 360$. Use the formula to:

a find the angle sum of a pentagon, which has 5 sides.

b show that the angle sum of a triangle is 180°.

HW HOMEWORK

3 SIMPLIFYING ALGEBRAIC EXPRESSIONS

DO YOU 'GET' ALGEBRA YET? ONCE YOU KNOW IT, IT'S NOT TOO HARD. KEEP TRAINING.

Name:

Due date:

Parent's signature:

Part A	/ 8 marks
Part B	/ 8 marks
Part C	/ 8 marks
Part D	/ 8 marks
Total	/ 32 marks

PART A: MENTAL MATHS

Calculators not allowed

1 Evaluate each expression.

a $120 - 4 \times 8$ ______

b $32.6 \div 0.4$

______

______

c 20% of 45 ______

2 Solve the equation $4n + 5 = 7$.

______

______

3 This equilateral triangle has a perimeter of 24 cm. What is the length of each side?

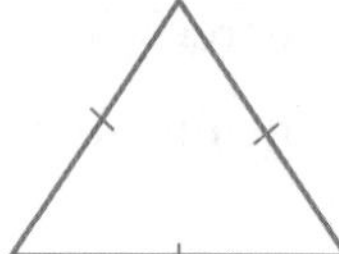

4 Complete: $3 : 5 =$ ______ $: 25$

5 Find the lowest common multiple of 8 and 10.

______

6 List the sample space when a die is rolled.

______

PART B: REVIEW

1 Write each statement as an algebraic expression using n to stand for the number.

a 4 times the number, then add 10

______

b the previous odd number if n is odd

______

c the difference between 5 and the number multiplied by itself

______

2 Simplify each expression.

a $7 \times p \times 4 \times d$

______

b $h + 2h + 3h$

______

c $(16 + r) \div 2$

______

3 If $x = -1$ and $y = -2$, find the value of:

a $2xy$

______

b $3x + 5y$

______

C S F

HW HOMEWORK

9780170454520

PART C: PRACTICE

› Adding and subtracting terms
› Multiplying and dividing terms

1 Simplify each expression.

a $3a + 4b - a - 2b$

b $5w - 2 + 6w - 3$

c $5a - 3b + 8b - 6a$

d $4mn - 7m^2 + 3nm - m^2$

e $5b \times 3c$

f $-5m \times (-2m)$

g $12ab \div 4b$

h $-\dfrac{36mn}{9n^2}$

PART D: NUMERACY AND LITERACY

1 Explain what *like terms* are, giving an example.

2 Explain why:

a $x + x = 2x$

b $\dfrac{x}{x} = 1$

3 Show by substituting a value for a that

$$5a - 4a = a.$$

4 Write as a simplified algebraic expression:

a the product of $8y$ and $4y$

b the sum of $8y$ and $4y$

c the quotient of $8y$ and $4y$

d the difference between $8y$ and $4y$

HW HOMEWORK

C S F

3 EXPANDING AND FACTORISING EXPRESSIONS

Name:

Due date:

Parent's signature:

Part A	/ 8 marks
Part B	/ 8 marks
Part C	/ 8 marks
Part D	/ 8 marks
Total	/ 32 marks

EXPANDING AND FACTORISING ARE THE HARDEST PARTS OF THIS TOPIC. ARE YOU MASTERING ALGEBRA?

PART A: MENTAL MATHS

Calculators not allowed

1 Complete this number pattern:

16, 22, 28, 34, ______

2 Evaluate each expression.

a $84 - 5 \times 7$

______

b 0.07×0.3

______

3 Mark a pair of alternate angles.

4 Simplify 56 : 21. ______

5 If $m = -5$, evaluate $7m + 18$. ______

6 Write 0.125 as a simple fraction. ______

7 Find the area of this triangle.

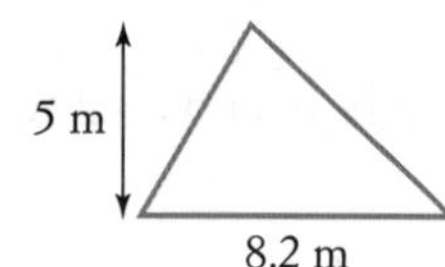

______

PART B: REVIEW

1 Simplify each expression.

a $3d + 15 + 2d - 28$

______

b $10r - 5 - 6r + 9$

______

c $6 \times 4n$

______

d $5 \times (-5c)$

______

2 Evaluate each expression.

a $4 \times (7 + 3)$

______

b $4 \times 7 + 4 \times 3$

______

3 Find the highest common factor of:

a 18 and 24

______

b 35 and 28

______

C S F

HW HOMEWORK

9780170454520

PART C: PRACTICE

› Expanding expressions
› Factorising expressions
› Factorising with negative terms

1 Expand each expression.

a $5(m - 4)$

b $-7(a + 6)$

c $4(2a - 3)$

2 Find the highest common factor of:

a $36y$ and $15y$

b $12a$ and $16ab$

3 Factorise each expression.

a $12a + 16ab$

b $10b - 4b^2$

c $-18mn - 9np$

C
S
F

PART D: NUMERACY AND LITERACY

1 Expand and simplify each expression.

a $4(m + 4) - 3(m - 5)$.

b $e(e + 15) + e(e + 7)$.

2 Explain why $20a - 15ab = 5(4a - 3ab)$ is not factorised completely and write the correct factorisation.

3 **a** Complete: The distributive law is:

$a(b + c) =$ ____________

b Use the distributive law to evaluate 27×11 mentally.

4 Complete: To factorise completely, you must take out the ________ common ________ in front of the brackets.

5 Factorise $8st + 24rs - 16sv$.

6 Show that $7m - 14 = 7(m - 2)$ by substituting $m = 5$.

HW HOMEWORK

3 ALGEBRA REVIEW

WE'RE UP TO THE END OF THE ALGEBRA TOPIC NOW. HOPE YOU'VE HAD A GOOD WORKOUT.

Name:

Due date:

Parent's signature:

Part A	/ 8 marks
Part B	/ 8 marks
Part C	/ 8 marks
Part D	/ 8 marks
Total	/ 32 marks

PART A: MENTAL MATHS

Calculators not allowed

1 Evaluate $160 - 96 \div 8$.

______

2 A traffic light shows red for 75 seconds, green for 43 seconds and amber for 2 seconds.
What is the probability that a car arriving at a random time faces a red light?

______

3 Draw an obtuse-angled triangle.

4 Write 6.05 a.m. in 24-hour time. ______

5 Convert 0.375 to:

a a percentage ______

b a simple fraction ______

6 Solve the equation $\frac{n+4}{2} = 5$.

______

7 If this triangle has an area of 27 m^2, what is its height?

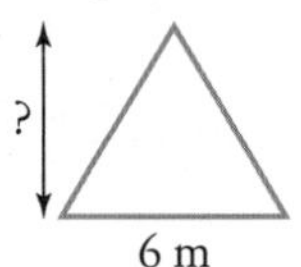

______

PART B: REVIEW

1 Simplify each expression.

a $5w + 4v - 3w - v$

______

b $\frac{52ab}{4b}$

______

c $3s \times 5t \times (-2v)$

______

d $p \div 4 - r \times r$

______

2 Expand each expression.

a $9(w + 5)$

______

b $m(3n - 2)$

______

3 If x and y are any 2 numbers, write an algebraic expression for:

a the difference between triple x and 5 times y

______

b the average of x and y

______

C S F

9780170454520

PART C: PRACTICE

› Algebra revision

1 Write a simplified algebraic expression for the perimeter of this rectangle.

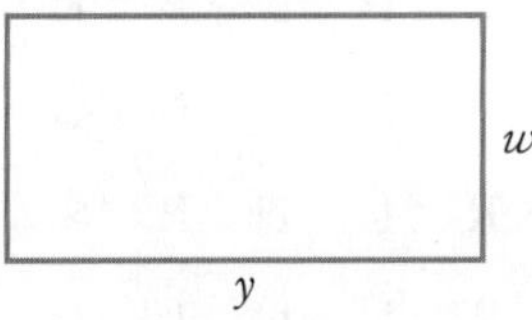

____

2 If $a = -4$ and $b = 5$, evaluate:

a $2a - 15$

____

b $a^2 + 7b$

____

3 Find the highest common factor of $15aw$ and $12ab$.

____

4 Factorise each expression.

a $12abc + 48bd$

____

b $-15aw - 10ab$

____

5 Expand and simplify each expression.

a $20 - 4(3f + 5)$

____

b $2(m - 6) + 7(2m - 1)$

____

C
S
F

PART D: NUMERACY AND LITERACY

1 **a** Write in words the meaning of $3(n + 4)$.

____

____

____

b Hence show why $3(n + 4) = 3n + 12$.

____

____

____

2 Show that $7z - 5z = 2z$ by substituting a value for z.

____

____

3 (2 marks) Write an algebraic expression for the number of times $2d$ divides into $18a^2d$ and then simplify the expression.

____

4 **a** Complete this distributive law:

$a(b - c) =$ ______________________

b Use the distributive law to evaluate 16×9.

____

____

5 The formula for converting Celsius temperatures to Fahrenheit temperatures is $F = \frac{9}{5}C + 32$. Use it to convert 18°C to Fahrenheit.

____

____

HW HOMEWORK

3 ALGEBRA FIND-A-WORD

MY STRATEGY WITH THESE IS TO CHOOSE AN UNUSUAL LETTER IN EACH WORD AND TO LOOK FOR THAT LETTER IN THE PUZZLE.

H I B T C A R T B U S T C U A C F C H F P K N T S

F B E R M C E X E A N F C B R O T A N I M O N E D

R R W C A C T A B L E N S U U W M R L H P S O I B

Q H A Z Z C G W D O G I C D D D J S U U L D X E D

L W C C C E K C S U A C F E T O O R G Q D F T J I

X D F E T V X E I F T P V R N O R A L A V A A M F

P O K E X I E P T N I K A E E C U P F J U G L V F

L I F R T D O B A S V P R V L S D N G L S F G E E

L A U J I U X N Q N E F I I A K F X A G M A E B R

Q L R V B Q T E I T D B A T V X T V J M Y C B A E

E Z I E T G S I X P V W B I I C E E T U L T R O N

V D P Z M D I Y T P E B L S U S U S E S W O A S C

T J I S E U E D M S R V E O Q Q L A R W F R A N E

Z C J C E D N C C B B E N P E J I E M X A I G N G

D P P J H W A O R P O U S Q Q Q R R W M H S Y H Q

T N E I T O U Q R E D L S S R O T C A F X E B Q H

X G N I P U O R G P A M S R I O Z N Z D F I R S Q

N S X J M O S E D V C S H U S O C I Z P W E N G J

J P A T T E R N J L F A E L L Q N F O R M U L A J

E U C X I M U L T I P L Y V Y E O S I M P L I F Y

Find these words in the puzzle above.

ADD	SYMBOLS	TABLE	TERM
DENOMINATOR	ALGEBRA	BRACKETS	DECREASE
EVALUATE	DIFFERENCE	DIVIDE	EQUIVALENT
FACTORISE	EXPAND	EXPRESSION	FACTOR
GROUPING	FORMULA	FRACTION	LIKE
HCF	INCREASE	LCM	POSITIVE
MULTIPLY	NEGATIVE	PATTERN	RULE
PRODUCT	PRONUMERAL	QUOTIENT	SUM
SIMPLIFY	SUBSTITUTE	SUBTRACT	VARIABLE

STARTUP ASSIGNMENT 4

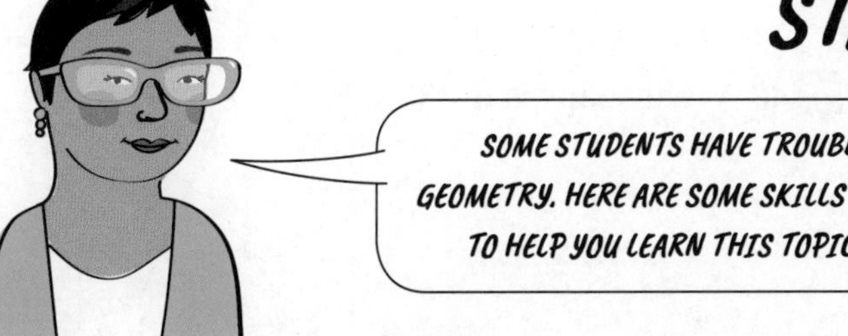

SOME STUDENTS HAVE TROUBLE WITH GEOMETRY. HERE ARE SOME SKILLS TO PRACTISE TO HELP YOU LEARN THIS TOPIC BETTER.

PART A: BASIC SKILLS

/ 15 marks

1 Convert 2.2 hours to hours and minutes.

______

2 Simplify $-4 + 5 \times (-3)$. ______

3 What is the reciprocal of $\frac{3}{5}$? ______

4 Is 29 prime or composite? ______

5 Expand $-6(3a - 2)$. ______

6 Evaluate $\sqrt{441}$ ______

7 If this triangle has an area of 35 cm^2, find b.

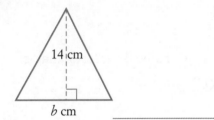

14 cm

b cm

______

8 **a** Draw a rhombus.

b How many axes of symmetry has a rhombus?

______

9 If $y = 4x + 5$, find y when $x = 3$.

______

10 Write these integers in ascending order:

$-3, 0, -5, 5, 1$. ______

11 Evaluate $8.43 \div 1.2$.

______

12 Simplify $4b \times (-3d) \times 2b$. ______

13 How can you test whether a number is divisible by 9?

______

14 Evaluate $\frac{4}{5} - \frac{1}{2}$.

______

PART B: GEOMETRY

/ 25 marks

15 How many degrees has a straight angle?

______

16 Mark the angle that is co-interior to x.

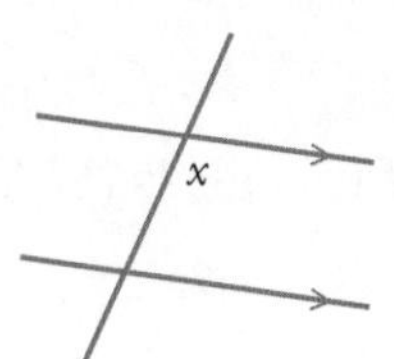

17 Draw and mark a reflex angle.

18 Are angles a and b supplementary or complementary? ______

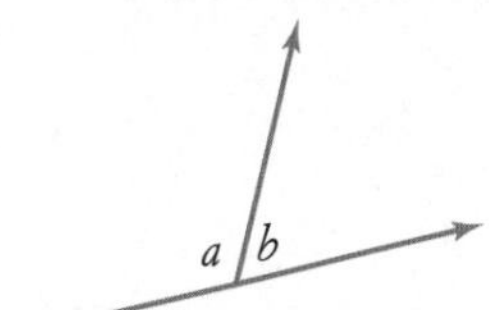

19 (2 marks) Find r and s.

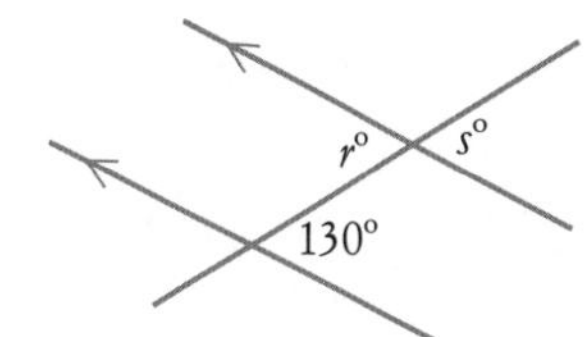

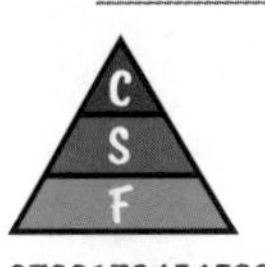

20 Draw the axes of symmetry of this square.

21 Which quadrilateral has one pair of parallel sides? ____________

22 **a** What type of angles are 105° and $p°$ in the diagram?

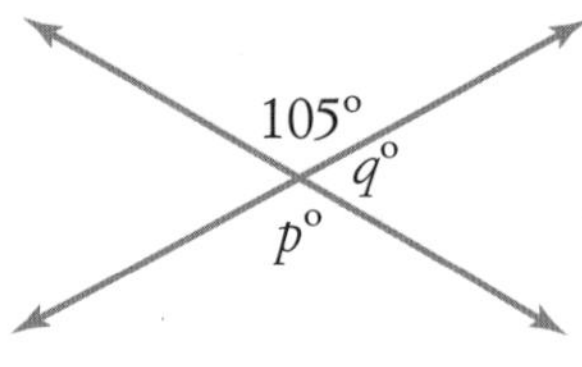

____

b (2 marks) Find p and q.

____

23 How many degrees in a revolution?

____

24 **a** What type of triangle is $\triangle PQR$?

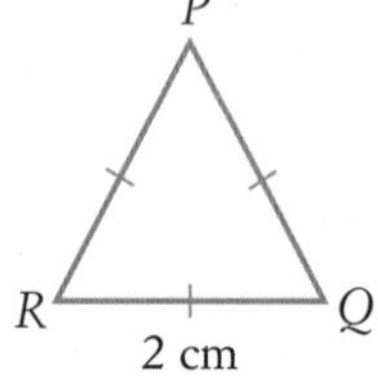

____

b What is the size of each angle in $\triangle PQR$?

____

25 Draw an obtuse-angled triangle.

26 Name any quadrilateral whose diagonals cross at right angles. ____________

27 **a** What shape is this?

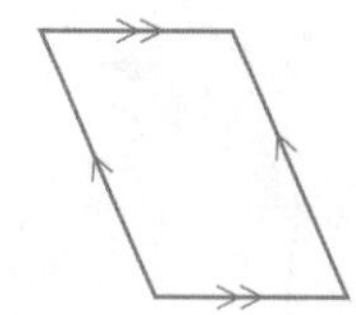

____

b How many axes of symmetry has it?

____

c What is its order of rotational symmetry?

____

28 What is the sum of the angles in a trapezium?

____

29 (2 marks) Sketch and label $\triangle BCD$, where $\angle B = 40°$, $\angle C = 90°$, $BC = 3$ cm.

30 **a** If *PQRS* is a kite, name the interval that is equal to *PQ*.

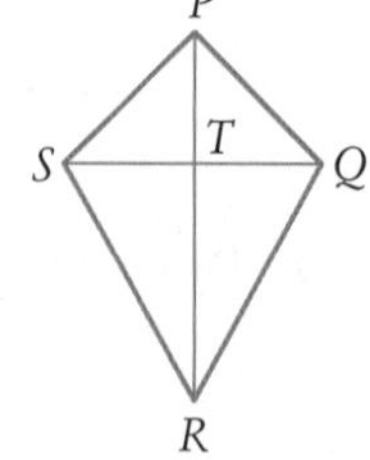

____

b Name the interval that is perpendicular to *PR*. ____________

c Name the angle that is equal to $\angle PSR$.

____

PART C: CHALLENGE

Bonus / 3 marks

Without lifting your pen, draw 4 straight lines that pass through every dot.

● ● ●

● ● ●

● ● ●

C S F

9780170454520

CLASSIFYING TRIANGLES 4

HERE'S A 3-PAGE PUZZLE THAT COVERS THE 6 DIFFERENT TYPES OF TRIANGLES.

Classifying triangles by their sides:

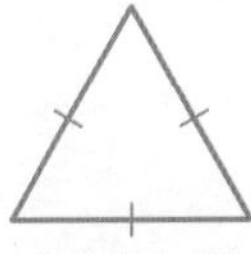
Equilateral

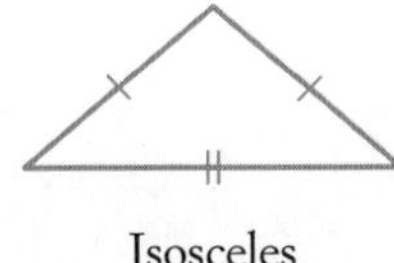
Isosceles

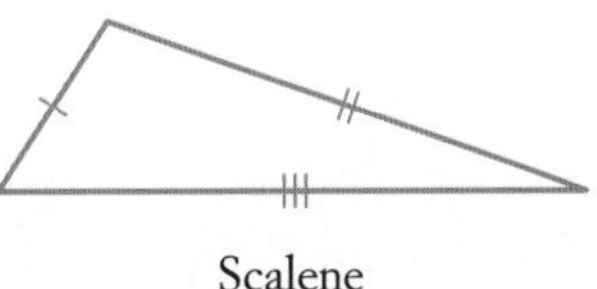
Scalene

Classifying triangles by their angles:

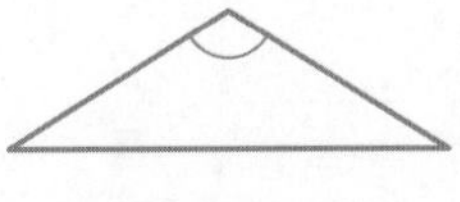
Obtuse-angled
(one angle > 90°)

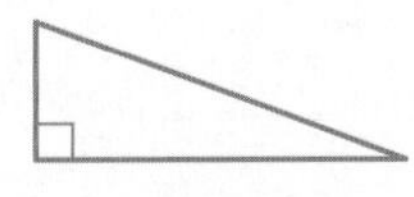
Right-angled
(one angle = 90°)

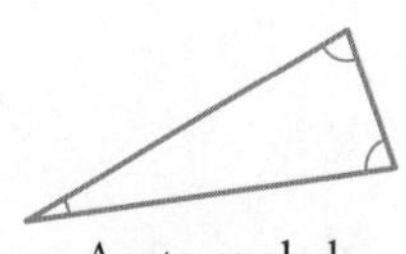
Acute-angled
(all angles < 90°)

'Courage is the most important of all the virtues, because without courage...

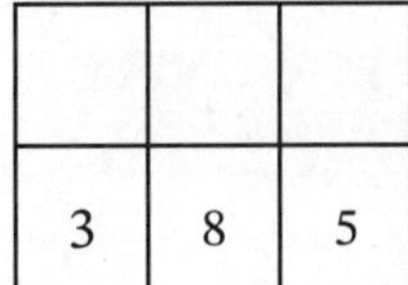

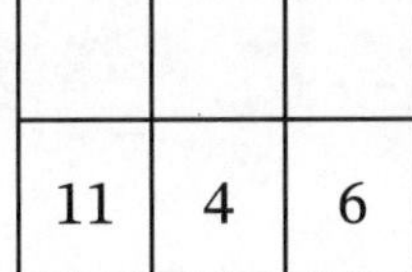

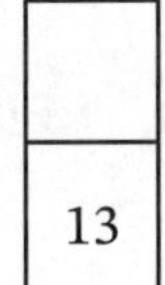

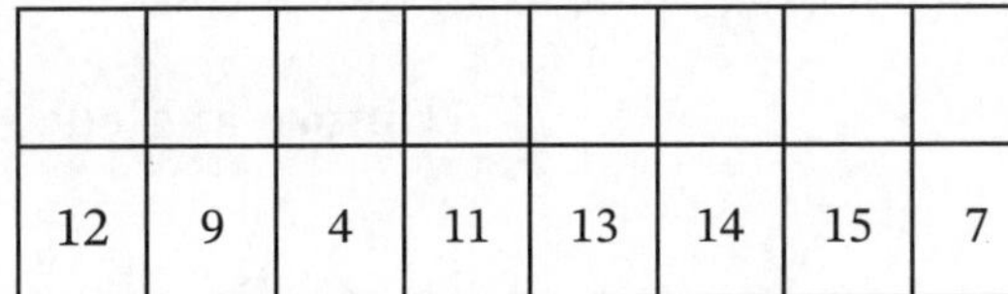

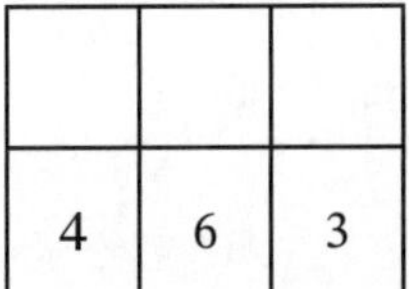

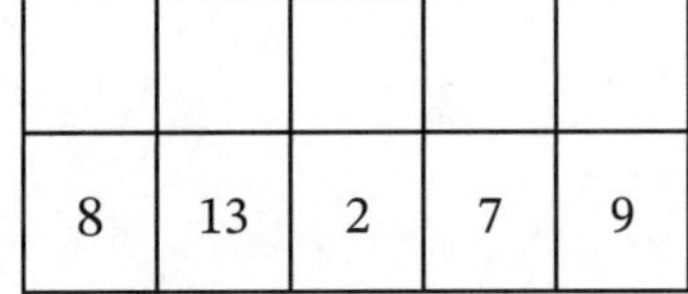

10	14	9	13	5	7

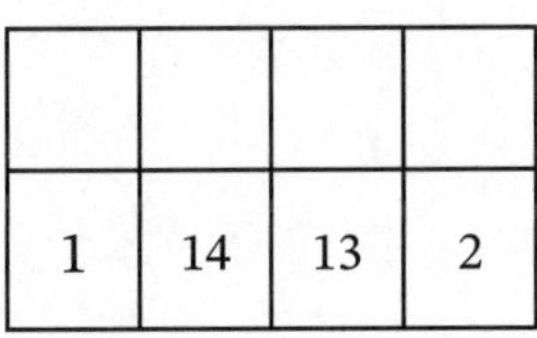

11	8	6	15	14	15	13	7	6	11	3

'.

The numbers in the grid above match the question numbers on the next 2 pages. Each question has a choice of 3 classifications, with 3 corresponding letters. Select the letter that goes with the correct classification for the triangle in each question and write the letters above the corresponding question numbers in the grid above to complete a quote from American author and poet Maya Angelou.

Classify the following triangles by their sides.

		Equilateral	Isosceles	Scalene
1	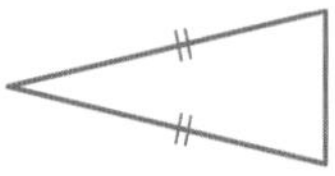	F	W	K
2	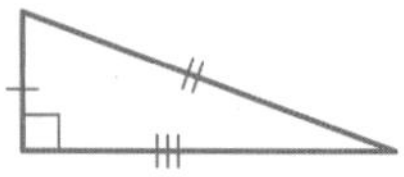	G	I	H
3		Y	T	S
4	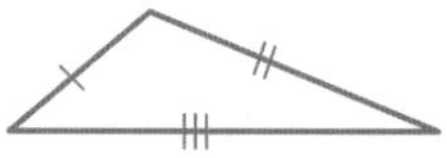	C	L	A
5	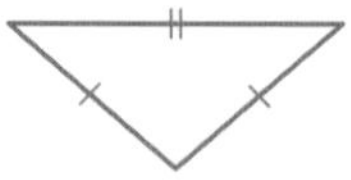	M	U	O

Classify the following triangles by their angles.

		Obtuse-angled	Right-angled	Acute-angled
6		D	F	N
7	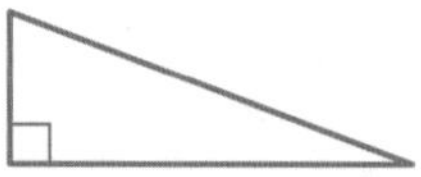	K	E	T
8		O	S	G
9	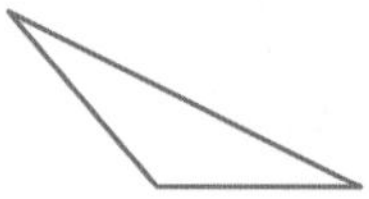	R	D	B
10		M	V	Q

PS PUZZLE SHEET

9780170454520

Classify the following triangles by side and angle.

	Isosceles obtuse-angled	Scalene right-angled	Isosceles right-angled	Scalene obtuse-angled	Equilateral acute-angled
11	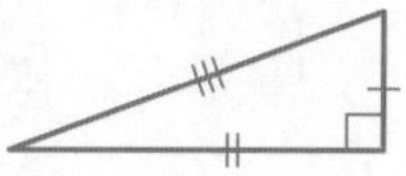 Y	C	X	R	T
12	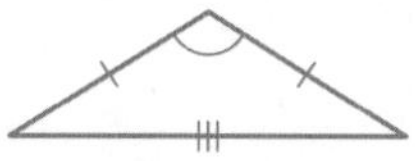 P	B	F	H	J
13	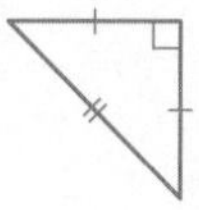 M	D	T	L	N
14	 G	K	Q	U	I
15	 A	H	O	S	W

PS PUZZLE SHEET

4 CLASSIFYING QUADRILATERALS

YOU NEED A RULER AND PROTRACTOR TO DRAW THESE QUADRILATERALS.

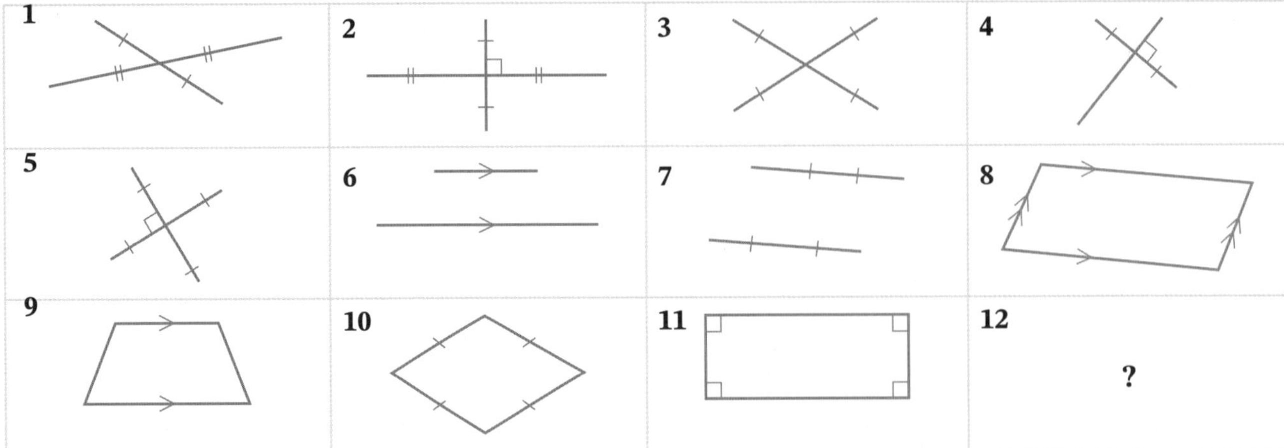

Construct each quadrilateral from the table by following its matching instructions below.

1 Draw 2 (unequal) diagonals that bisect each other. What quadrilateral is formed if you join the 4 ends? Complete: The diagonals of a ____________ bisect each other.

2 Draw 2 (unequal) diagonals that bisect each other *at right angles*. Complete: The diagonals of a ____________ bisect each other at right angles. Also, diagonals bisect the angles through which they pass. Mark this on your diagram.

3 Draw 2 equal diagonals that bisect each other. The diagonals of a ____________ are equal and bisect each other.

4 Draw 2 diagonals where only one diagonal is bisected by the other at right angles. One diagonal of a ____________ bisects the other at right angles. Mark where this diagonal bisects the angles it passes.

5 Draw 2 equal diagonals that bisect each other at right angles. The diagonals of a ____________ are equal and bisect each other at right angles. Mark the angles that are 45°.

6 Draw 2 (unequal) parallel sides and join the ends to make a quadrilateral. What shape is this?

____

7 Draw 2 *equal* parallel sides and join the ends. What shape is this? ____________

8 Draw a **parallelogram** (any shape with 2 pairs of parallel sides). Write True or False for each statement.

a A rectangle is a parallelogram ____________

b A square is a parallelogram ____________

c A trapezium is a parallelogram ____________

d A rhombus is a parallelogram ____________

e A kite is a parallelogram ____________

9 Draw a **trapezium** (any quadrilateral with [at least] one pair of parallel sides).
Write True or False for each statement.

a A rectangle is a trapezium ____________

b A square is a trapezium ____________

c A parallelogram is a trapezium ____________

d A rhombus is a trapezium ____________

e A kite is a trapezium ____________

10 Draw a **rhombus** (any shape with 4 equal sides). True or False?

a A kite is a rhombus ____________

b A square is a rhombus ____________

c A parallelogram is a rhombus ____________

d A rectangle is a rhombus ____________

11 Draw a **rectangle** (any shape with 4 right angles). True or False?

a A kite is a rectangle ____________

b A parallelogram is a rectangle ____________

c A rhombus is a rectangle ____________

d A square is a rectangle ____________

12 Write the name of the *most general quadrilateral* in which:

a all sides are equal ____________

b opposite sides are equal ____________

c opposite angles are equal ____________

d diagonals bisect each other ____________

e opposite sides are parallel ____________

f diagonals bisect at 90° ____________

g diagonals are equal and bisect each other ____________

h all angles are 90° ____________

i one pair of opposite sides are parallel *and* equal ____________

j diagonals cross at 90° ____________

Challenge questions: True or false?

a A rhombus is a square. ____________

b A rhombus is a kite. ____________

4 FIND THE UNKNOWN ANGLE

FOR QUESTION 9, YOU MIGHT WANT TO DRAW ANOTHER PARALLEL LINE.

Find the value of the variable(s) in each diagram, giving reasons.

1

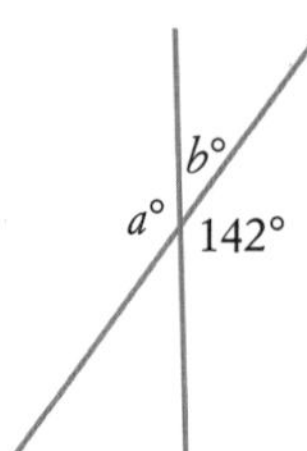

2

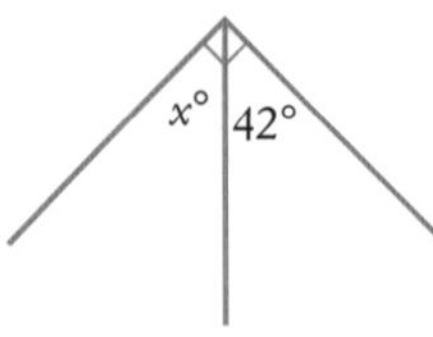

3

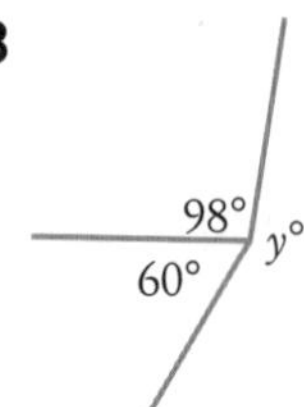

4

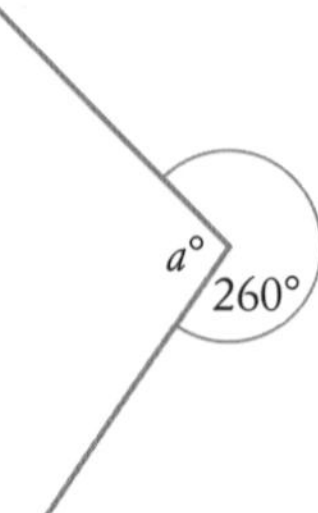

5

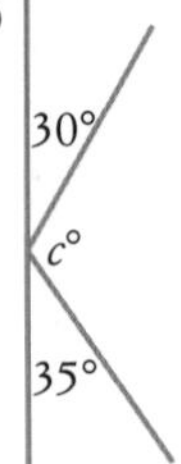

6

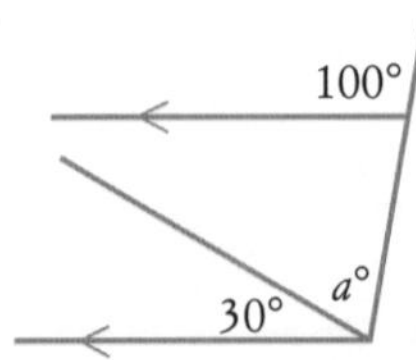

7

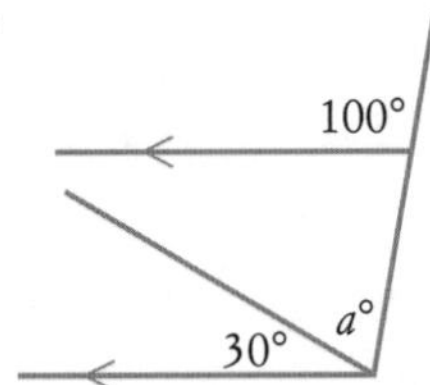

8

9

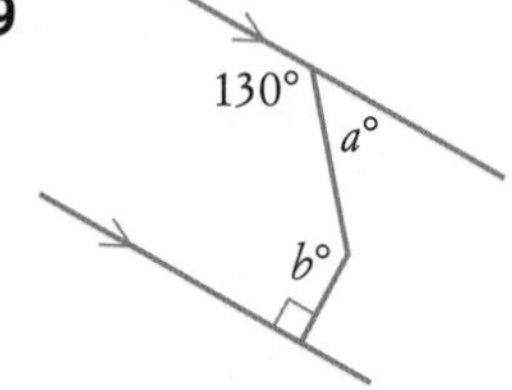

10

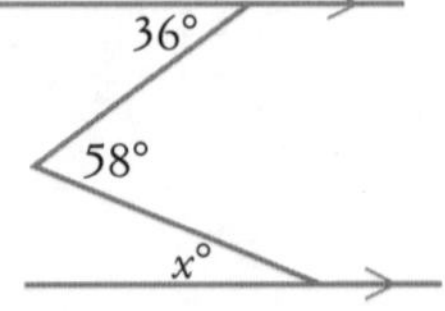

11

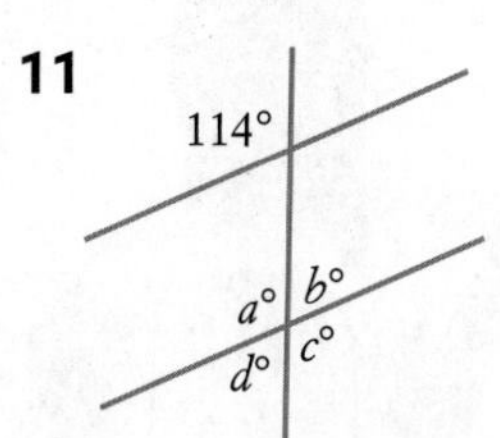
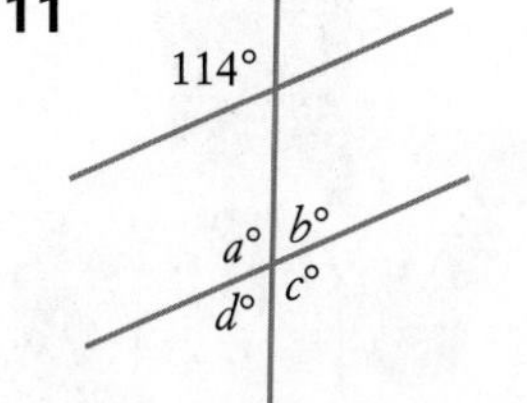

12

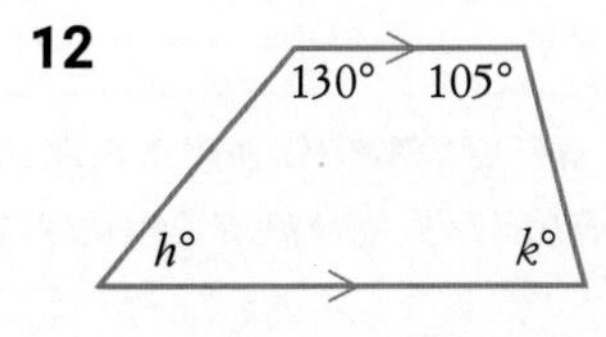

13

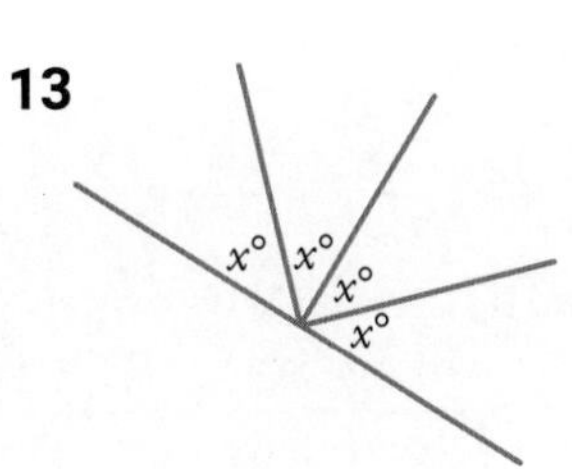

14

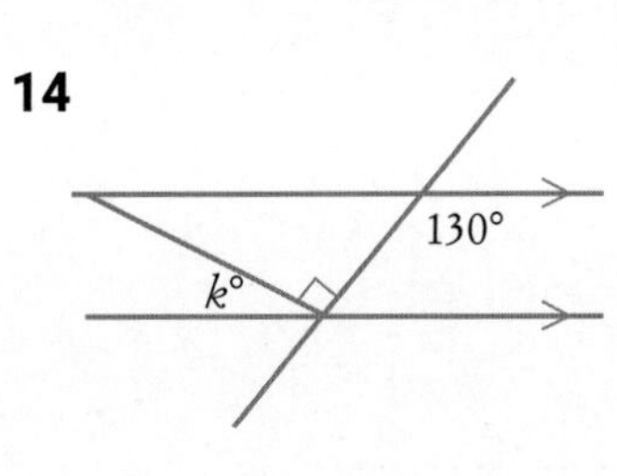

15

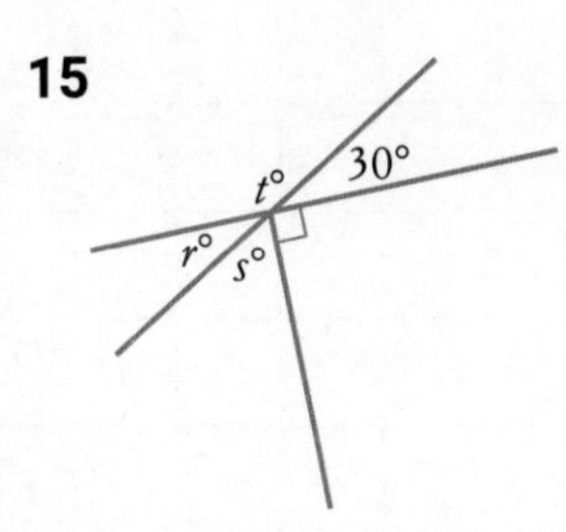

16

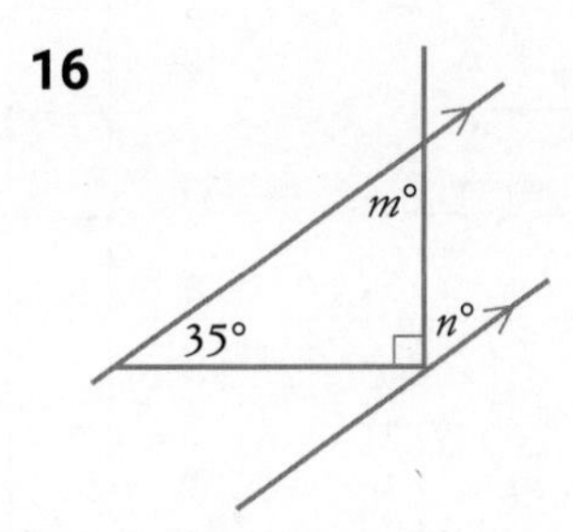

17

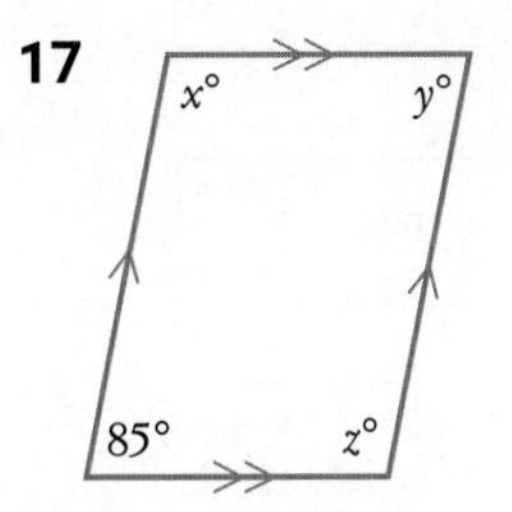

18

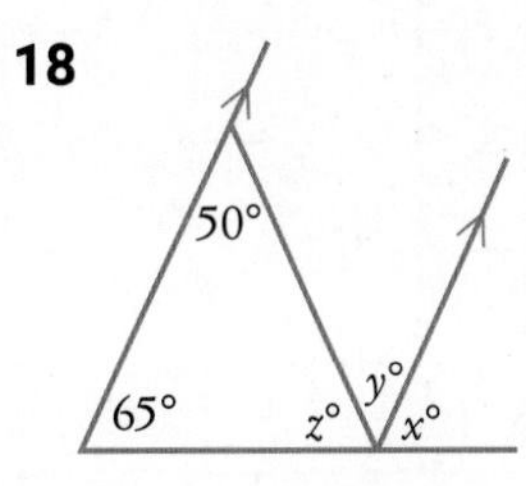

19

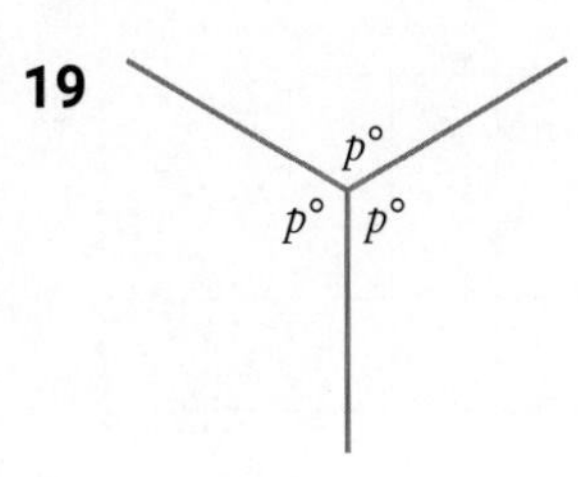

20

x°
x°
100°

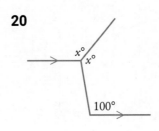

4 GEOMETRY CROSSWORD

MATHS TERMINOLOGY IS ESPECIALLY IMPORTANT IN GEOMETRY. HOW WELL DO YOU KNOW YOUR KEYWORDS?

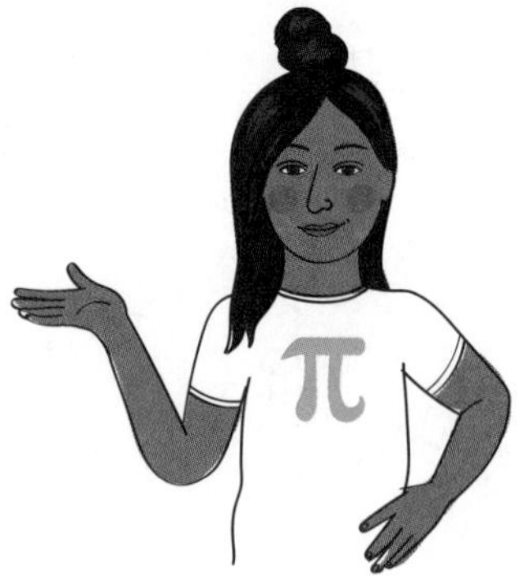

9780170454520

ACROSS

4 Quadrilateral with 4 equal angles

5 Quadrilateral with 4 equal sides

6 To cut in half

9 Matching angles on 2 lines crossed by a transversal

11 Outside angle formed by extending a side of a triangle

13 V-word meaning corner

14 Triangle with 2 equal sides

16 Less than a right angle

17 Line that crosses another line at right angles

20 'Together inside' angles between parallel lines cut by transversal

21 Between 90° and 180°

22 Equal angles on either side of transversal between parallel lines

23 Lines that never cross

24 A 90° angle

25 Co-interior angles on parallel lines are ________________

26 A measure of turning between 2 intersecting lines is called an ________________

27 Equal 'opposite' angles are formed when 2 lines cross

DOWN

1 Quadrilateral with one pair of parallel sides

2 A right-angled triangular ruler is called a ________ square.

3 Measuring and drawing stick

7 Geometrical word for 'draw'

8 Quadrilateral with 2 pairs of equal adjacent sides

9 Angles that add to 90°

10 Quadrilateral with opposite angles equal

12 Any 4-sided polygon

15 Sharp instruments for drawing circles

17 This instrument measures angle sizes

18 360° angle

19 All angles 60° in this triangle

25 The answer to an addition

4 ANGLE GEOMETRY

A LOT OF TERMINOLOGY IN THIS ASSIGNMENT AS WELL.

Name:

Due date:

Parent's signature:

Part A	/ 8 marks
Part B	/ 8 marks
Part C	/ 8 marks
Part D	/ 8 marks
Total	/ 32 marks

PART A: MENTAL MATHS

Calculators not allowed

1 Solve the equation $\frac{4w}{5} = 6$.

2 Write 18:35 in 12-hour time.

3 Simplify 60 : 48.

4 Round 4.568 to one decimal place.

5 What is the probability of a baby being born on a weekday rather than a weekend?

6 Find the perimeter of this trapezium.

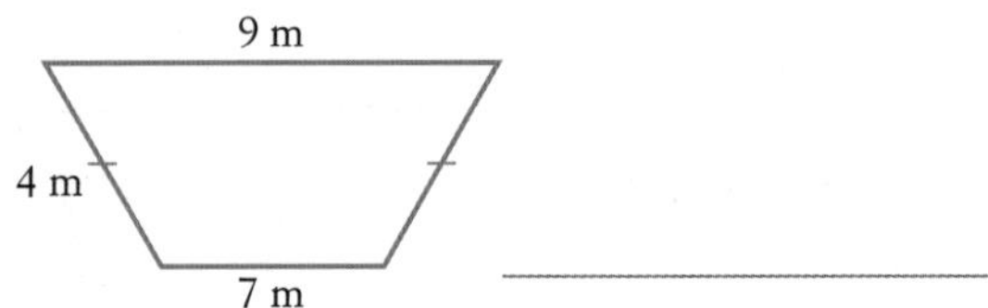

7 Evaluate 200×0.008.

8 Simplify $18mn - 3mn + 12mn$.

PART B: REVIEW

1 Name each type of angle drawn.

a 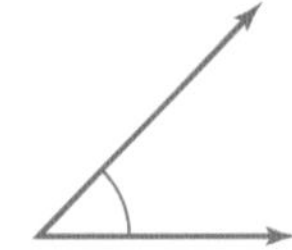

b

c 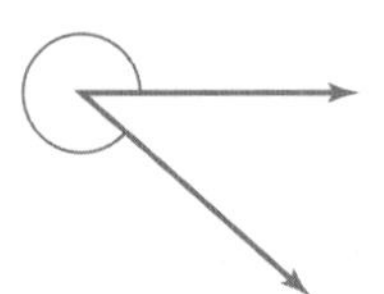

d 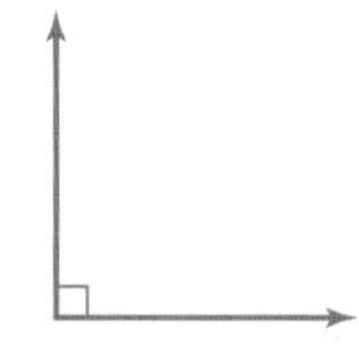

2 Draw:

a an obtuse angle

b a pair of vertically opposite angles.

3 Complete:

a The supplement of 48° is ______°

b An obtuse angle is between ______° and ______°

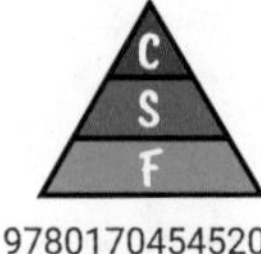

9780170454520

PART C: PRACTICE

› Angle geometry
› Angles on parallel lines

1 Find the value of each variable.

a

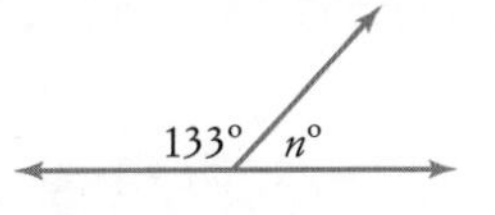

______

b

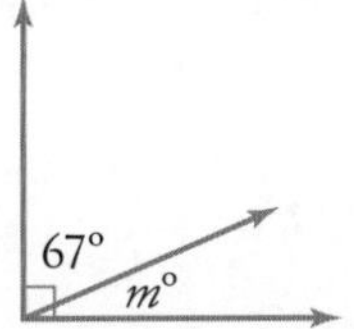

______

c

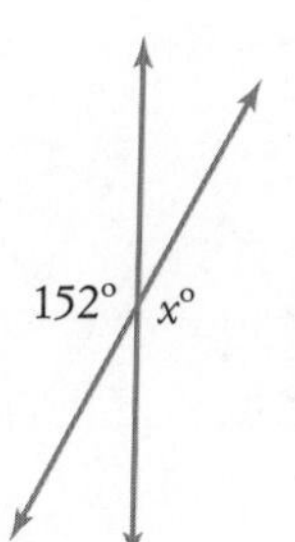

______

d

______

2 (4 marks) Find the value of each variable, giving a reason.

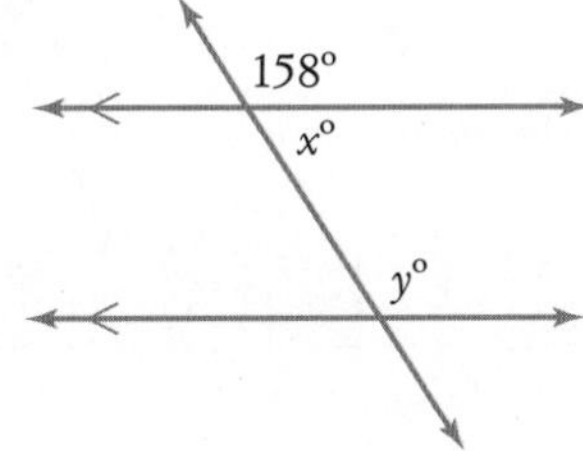

______

______

______

______

PART D: NUMERACY AND LITERACY

1 (2 marks) Complete:

a Angles at a point add up to ______°

b Angles on a straight line add up to ______°

2 What type of angles share a common arm?

______

3 (3 marks)

a Why are lines *AB* and *CD* parallel?

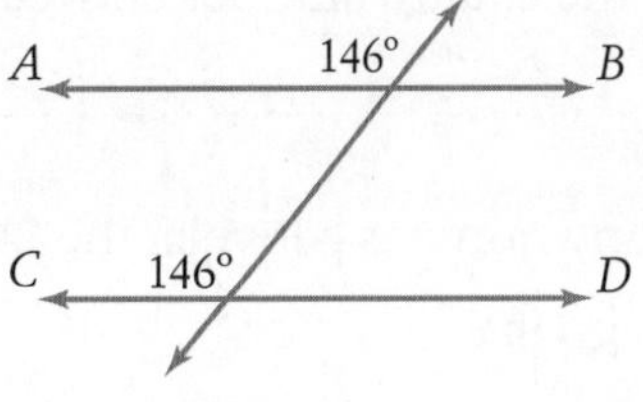

______

______

______

b What is the symbol for 'is parallel to'?

______

c Write on the diagram the sizes of the other 6 angles.

4 (2 marks) Complete: When parallel lines are crossed by a transversal:

a alternate angles are ______

b ______ angles are supplementary

HW HOMEWORK

C
S
F

4 SYMMETRY AND TRIANGLES

THIS ASSIGNMENT COVERS
LINE SYMMETRY AND
ROTATIONAL SYMMETRY.

Name:

Due date:

Parent's signature:

Part A	/ 8 marks
Part B	/ 8 marks
Part C	/ 8 marks
Part D	/ 8 marks
Total	/ 32 marks

PART A: MENTAL MATHS

Calculators not allowed

1 What is the square number between 60 and 70? ______

2 How many minutes between the 24-hour times 13:55 to 14:38?

______

3 Write a ratio that simplifies to 5 : 8. ______

4 Evaluate each expression.

a 27×8

______

b $\frac{180}{-6} \times \frac{-40}{-8}$

______

5 Find the area of this parallelogram.

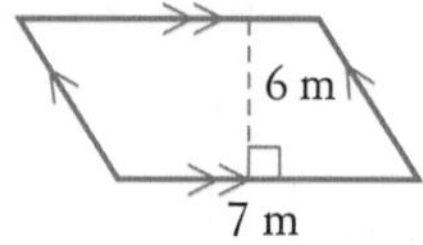

______

6 Write an algebraic expression for the average of k, b and x.

______

7 Simplify $3 \times a \times m \times a \times 2$. ______

PART B: REVIEW

1 Name each type of triangle.

a

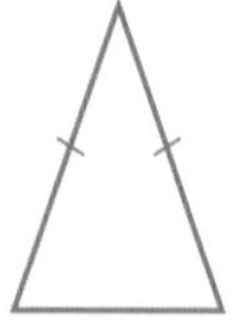

______

b

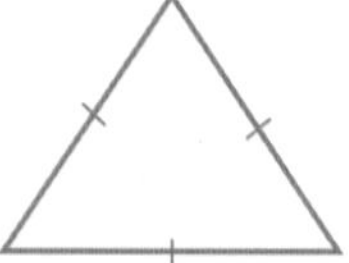

______

2 (4 marks) Find the value of each variable, giving a reason.

a

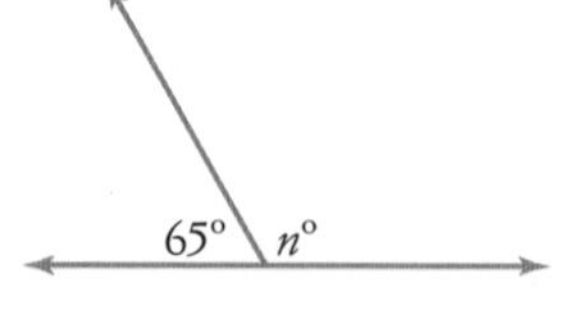

______

b

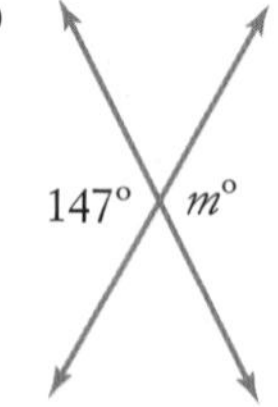

______

3 **a** Mark 2 co-interior angles on the diagram.

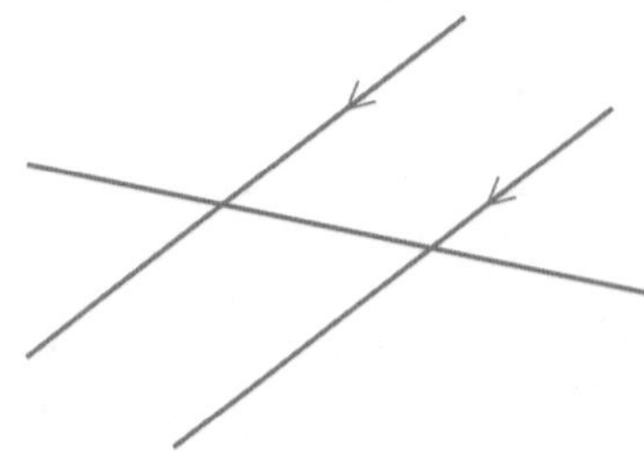

b Complete: Co-interior angles on parallel lines

______

C S F

9780170454520

HW HOMEWORK

PART C: REVIEW

› Line and rotational symmetry
› Classifying triangles

1 For each shape, draw its axes of symmetry.

a

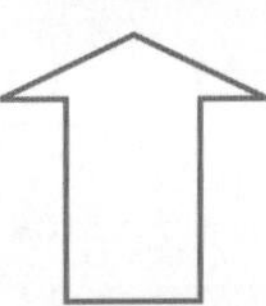

b

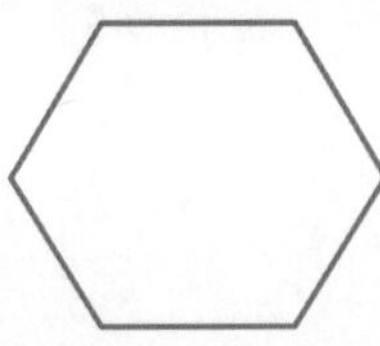

2 For each shape, state its order of rotational symmetry.

a

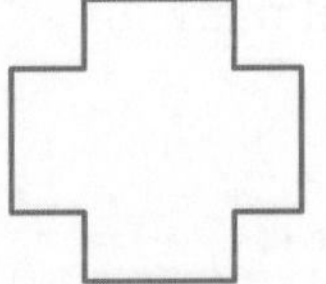

b

________________ ________________

3 What is the size of one angle in an equilateral triangle? ________________

4 Classify this triangle according to:

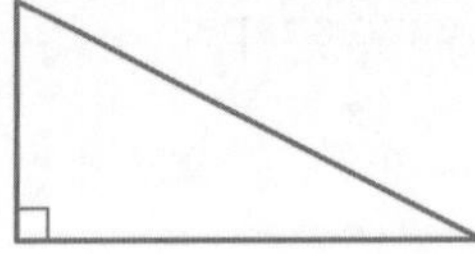

a its sides ________________

b its angles ________________

5 Draw an obtuse-angled isosceles triangle.

PART D: NUMERACY AND LITERACY

1 **a** What is an isosceles triangle?

____

____

b How many axes of symmetry has an isosceles triangle? ________________

c What order of rotational symmetry has an isosceles triangle? ________________

d How many equal angles has an isosceles triangle? ________________

2 What type of triangle has:

a no equal angles?

____

b 3 axes of symmetry?

____

c 2 angles below 90° and one angle above 90°?

____

3 Why is it impossible to have a triangle that is both right-angled and equilateral?

____

____

____

____

HW HOMEWORK

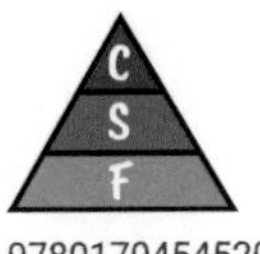

4 QUADRILATERALS AND ANGLE SUMS

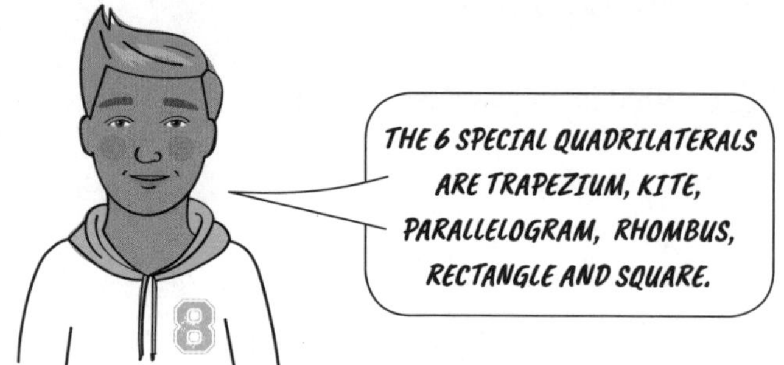

THE 6 SPECIAL QUADRILATERALS ARE TRAPEZIUM, KITE, PARALLELOGRAM, RHOMBUS, RECTANGLE AND SQUARE.

Name:

Due date:

Parent's signature:

Part A	/ 8 marks
Part B	/ 8 marks
Part C	/ 8 marks
Part D	/ 8 marks
Total	/ 32 marks

PPART A: MENTAL MATHS

Calculators not allowed

1 Simplify $4d + 1 - 3d - 9$. ______

2 Round 96.29 to the nearest tenth. ______

3 Evaluate each expression.

a $28 \div (-7) - (-5) \times 4$

______

b 2.5×0.02

______

4 Solve the equation $12 - 6n = 36$.

______

______

5 Write 9.50 p.m. in 24-hour time. ______

6 Simplify $\frac{27}{18}$. ______

7 Find the area of this triangle

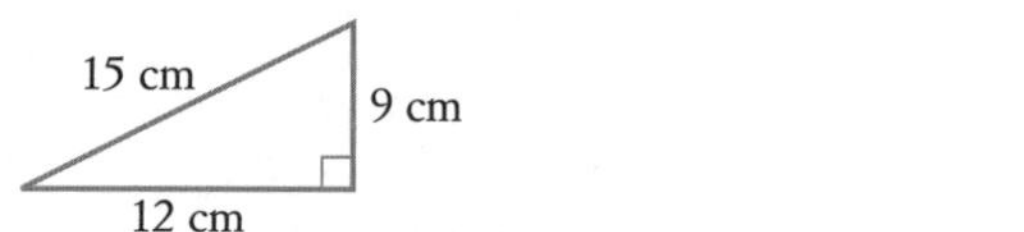

______

PART B: REVIEW

1 Draw an isosceles triangle and mark all equal angles.

2 Name this shape.

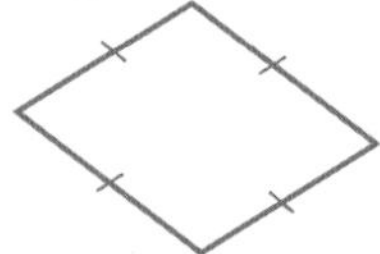

3 (4 marks) For each shape:

a

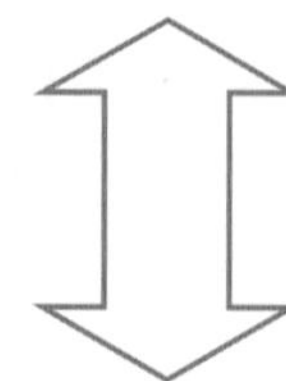

b

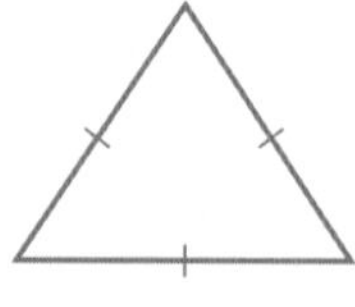

i count the number of axes of symmetry

______

ii state the order of rotational symmetry

______

4 Classify the triangle shown in question **3b** above according to:

a its sides

______

b its angles. ______

C S F

9780170454520

HW HOMEWORK

PART C: PRACTICE

› Classifying quadrilaterals
› Properties of quadrilaterals
› Angle sums of triangles and quadrilaterals

1 Name each quadrilateral.

a

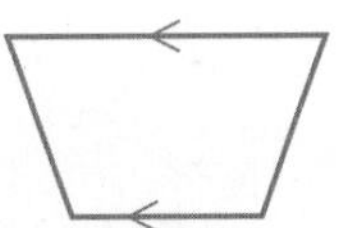

b

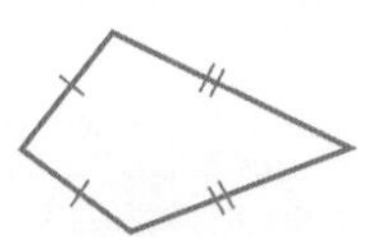

2 Complete:

a A rectangle is a quadrilateral with all angles ______

b A parallelogram is a quadrilateral with 2 pairs of ______

3 Find the value of each variable.

a

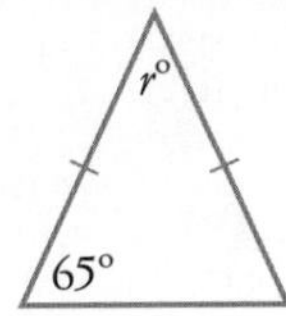

b

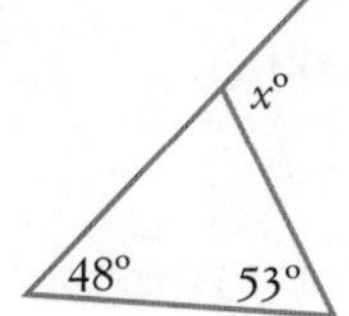

c

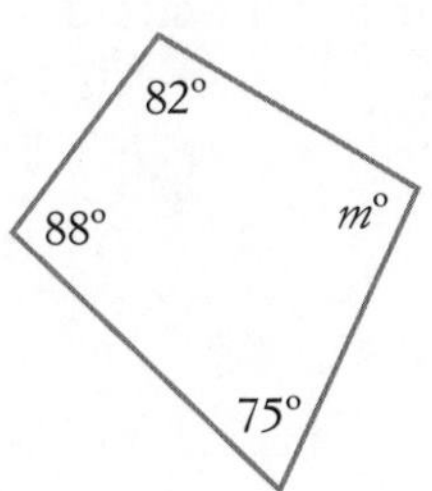

d

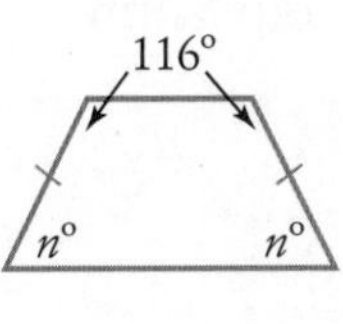

PART D: NUMERACY AND LITERACY

1 a Draw a rectangle and its diagonals.

b Write 2 properties about the diagonals of a rectangle.

2 a What is a rhombus?

b How does a square differ from a rhombus?

c If a rhombus has one angle 72°, what are the sizes of the other 3 angles?

3 Explain whether a square is a special type of parallelogram.

4 a What is the angle sum of a triangle?

b Why is the size of each angle in an equilateral triangle 60°?

HW HOMEWORK

C
S
F

WS WORKSHEET

5 STARTUP ASSIGNMENT 5

LET'S GET READY FOR THE AREA AND VOLUME TOPIC BY REVISING MEASUREMENT SKILLS.

PART A: BASIC SKILLS

/ 15 marks

1 Evaluate each expression.

a 48×9

______

b $14 + 6 \div 2 \times (-3)$

______

c $\frac{7}{9} - \frac{2}{3}$

______

d 2^5

______

2 Find u.

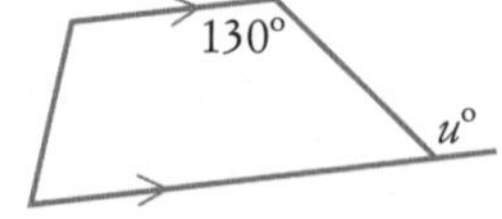

3 If p is an odd number, write an expression for the previous odd number. ______

4 What is the chance that a person chosen at random has a birthday in a month beginning with J? ______

5 An isosceles triangle has one angle of 130°. What are the sizes of the other 2 angles?

______

6 Simplify $9a + 7 - 4a - 10$.

______

7 Write $\frac{1}{3}$ as a decimal.

______

8 Use index notation to simplify $\frac{3^8}{3^4}$.

______

9 For this cube, find:

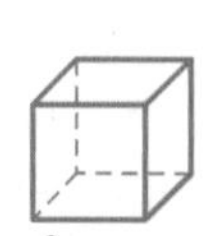

a the total area of its faces

______

b its volume ______

10 Factorise $12xy - 4y^2$. ______

11 True or false: The diagonals of a kite bisect each other. ______

PART B: AREA AND VOLUME

/ 25 marks

12 Draw:

a a trapezium **b** a semi-circle

13 For this square, find:

3 m

a its perimeter ______

b its area ______

14 Which solid shape has this net?

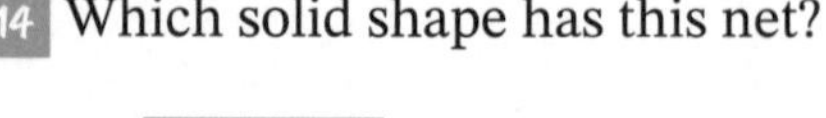

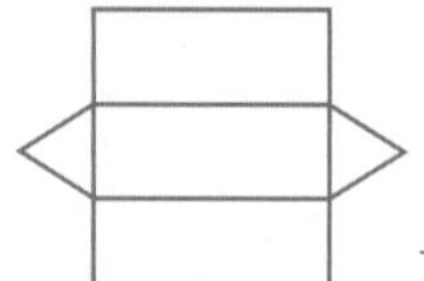

______

C S F

9780170454520

15 Complete:

a $1.8 \text{ m} =$ ______________________ mm

b $325 \text{ cm} =$ ______________________ m

c $96 \text{ cm} =$ ______________________ mm

d $2400 \text{ mL} =$ ______________________ L

e $1 \text{ ha} =$ ______________________ m^2

f $1 \text{ cm}^3 =$ ______________________ mL

16 How many degrees in a revolution?

____

17 For this triangle, find:

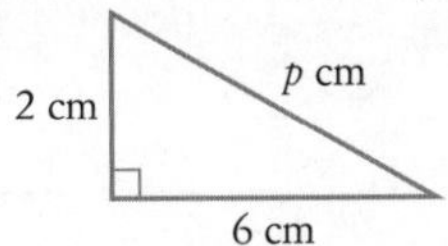

a p correct to 2 decimal places

____

b the area of the triangle

____

18 **a** Draw a rhombus and mark any axes of symmetry.

b What order of rotational symmetry has a rhombus? ______________________

19 Evaluate to 2 decimal places:

a $2 \times 3.1416 \times 7.2$ ______________________

b 3.1416×7.2^2 ______________________

20 For this shape, find:

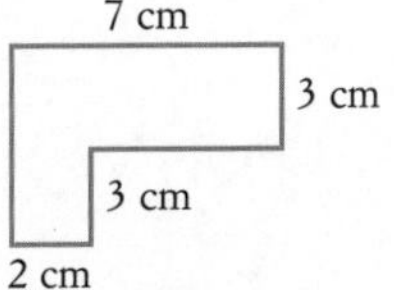

a its perimeter ______________________

b its area ______________________

21 Calculate the volume of this prism.

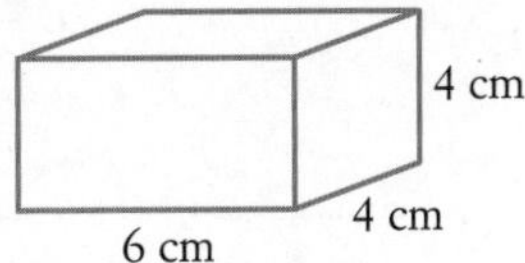

22 Find the area of this shape.

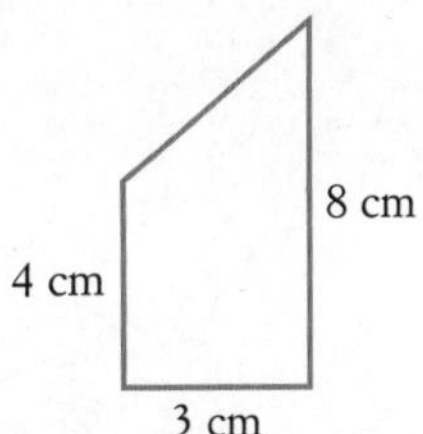

23 A square has an area of 4.84 cm^2. What is the length of one side?

____

24 For this rectangle, write a simplified expression for its:

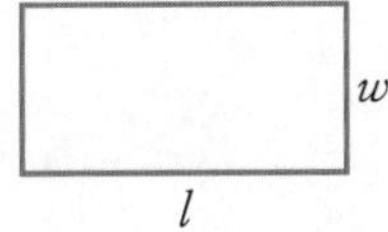

a perimeter ______________________

b area ______________________

PART C: CHALLENGE

Bonus / 3 marks

Draw a straight line across this clock face so that the numbers in each of the 2 parts have the same sum.

5 COMPOSITE AREAS

NOTICE THAT THESE AREA PROBLEMS RANGE FROM EASY TO HARD. HOW CAN YOU DIVIDE EACH SHAPE?

Find the shaded area of each figure.

1

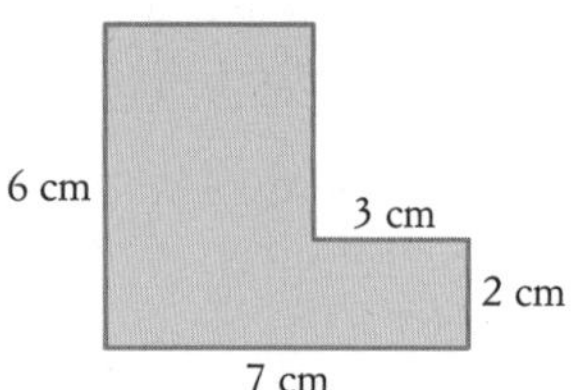

2

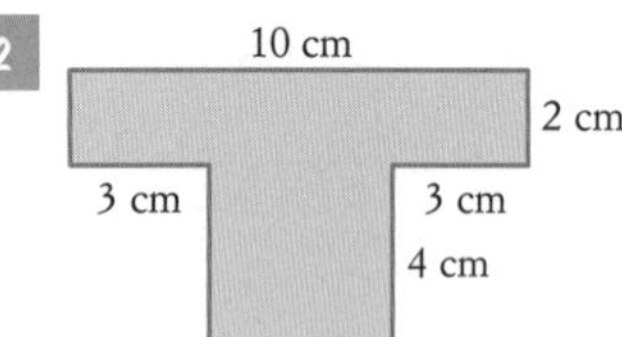

3

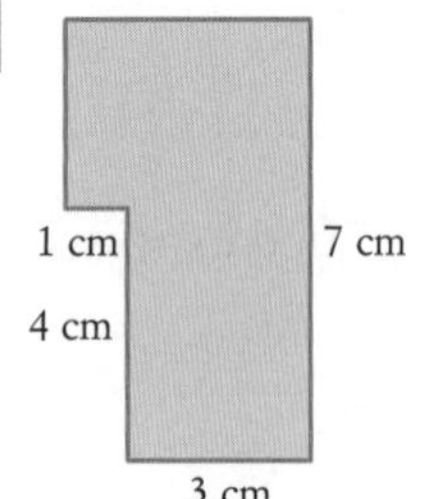

4

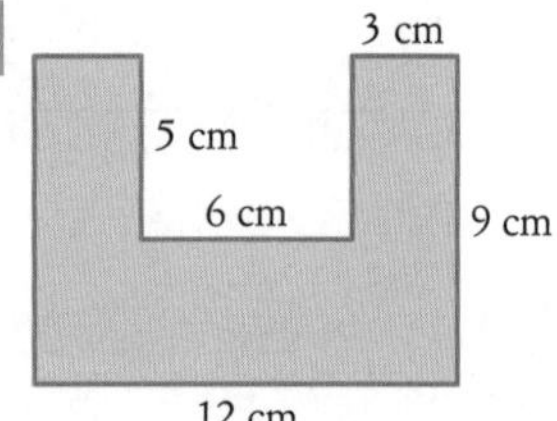

5

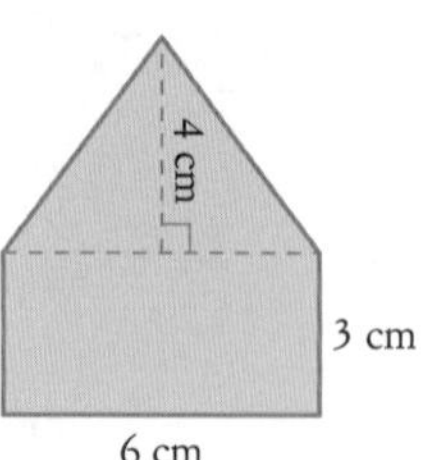

6

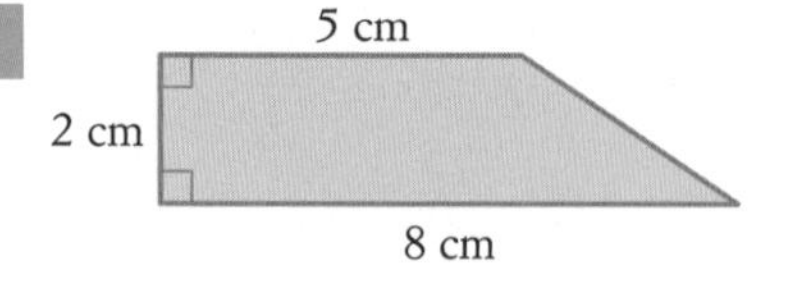

7

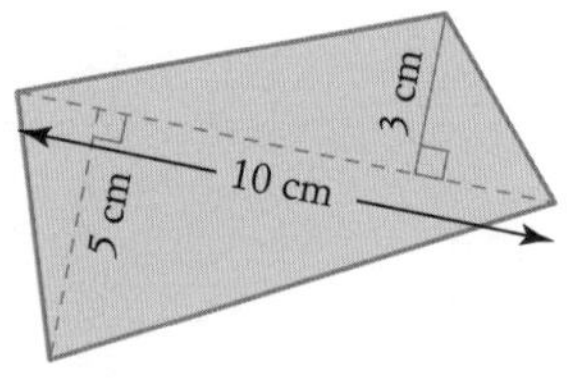

8

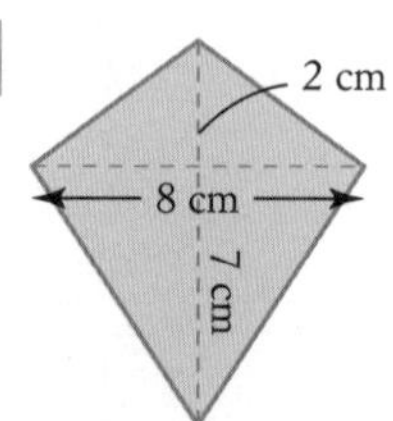

9

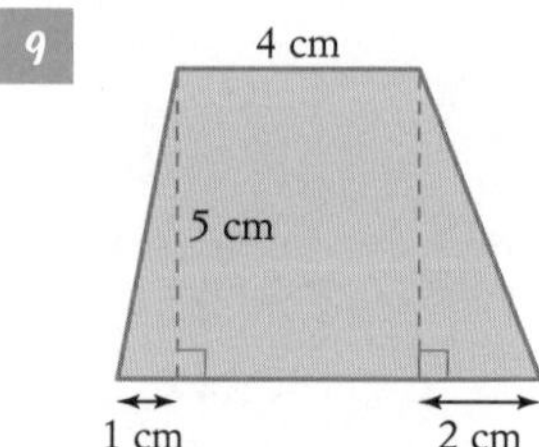

C
S
F

9780170454520

WS WORKSHEET

10

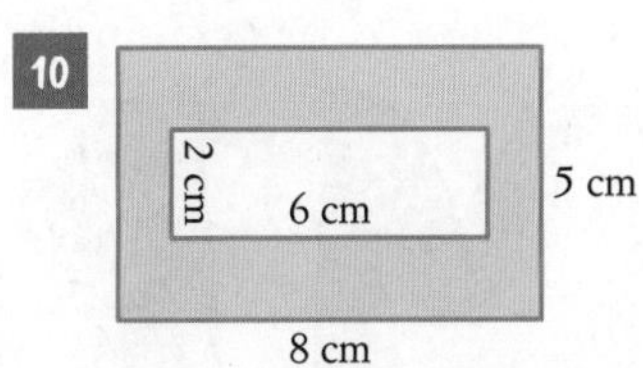

11

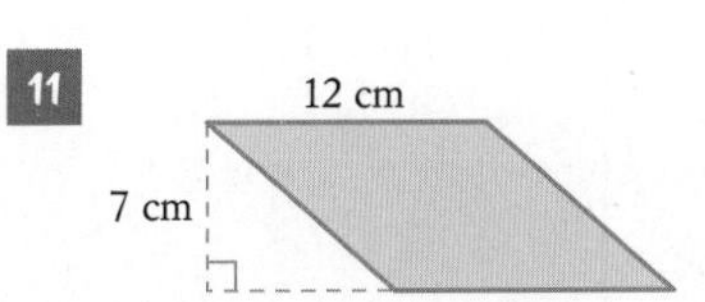

12

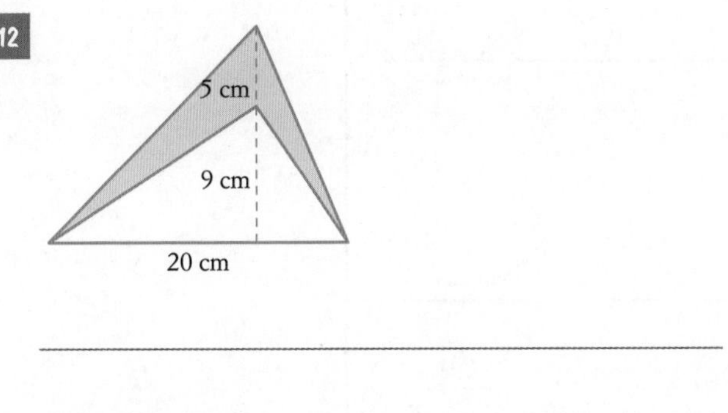

13

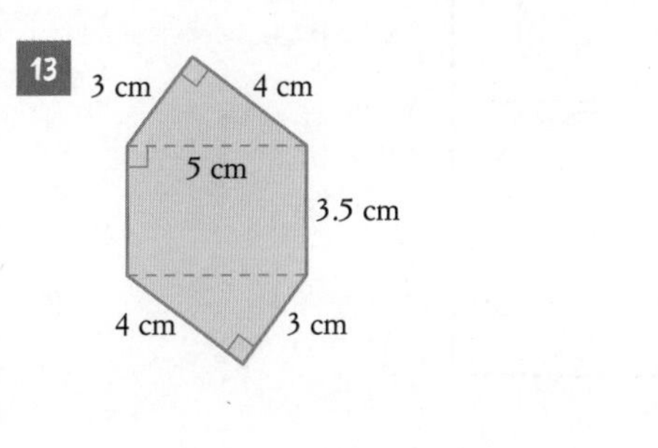

14

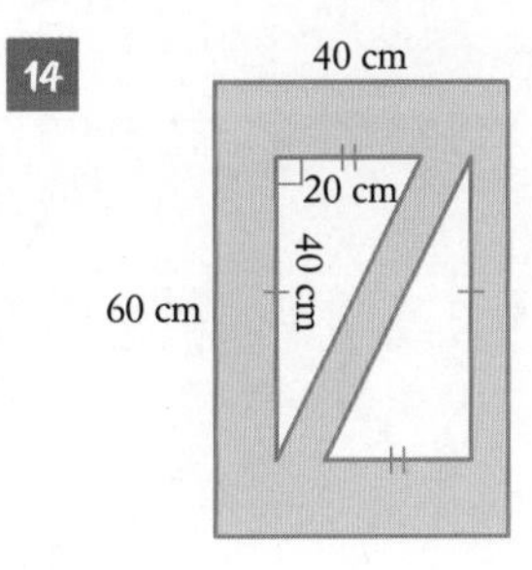

15

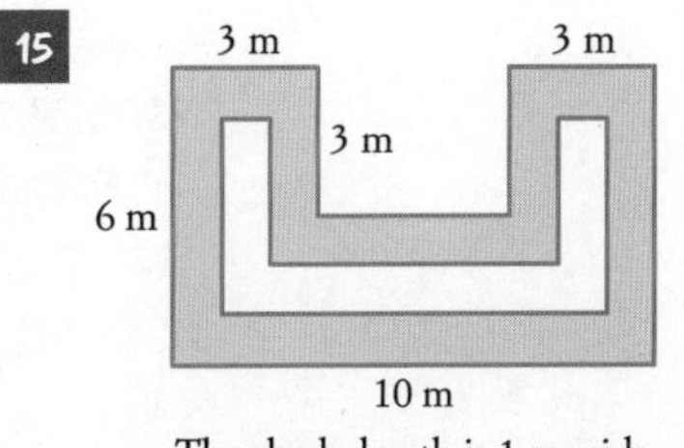

The shaded path is 1 m wide

16

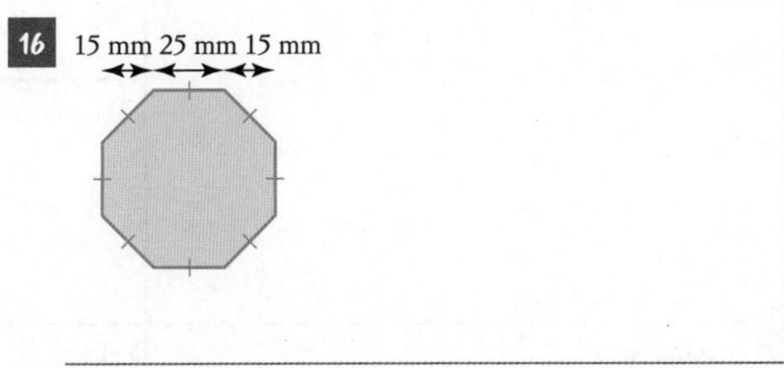

17

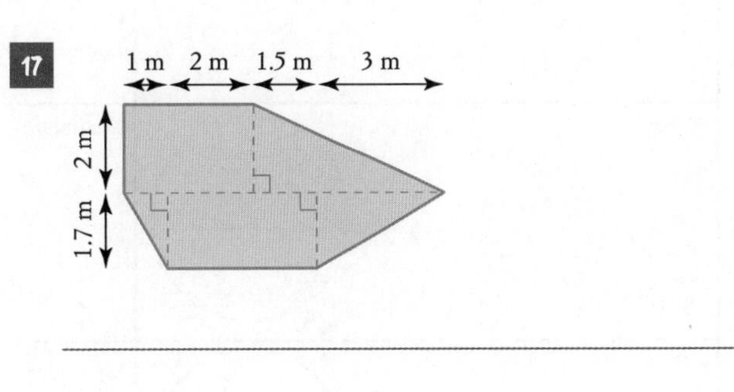

18

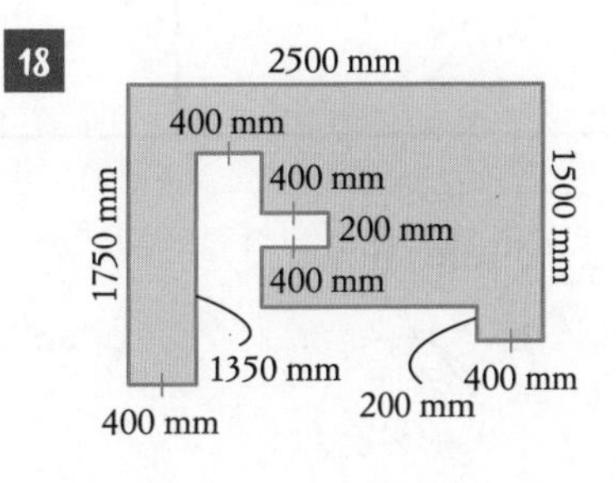

C S F

WS WORKSHEET

5 PARTS OF A CIRCLE

SECTOR GRAPH IS ANOTHER NAME FOR PIE CHART, JUST SAYIN'!

1 Draw and label each part of the circle named and write its meaning in your own words.

a centre: **b** radius: **c** diameter:	**g** quadrant: **h** semi-circle:
d circumference:	**i** chord:
e arc:	**j** segment:
f sector:	**k** tangent:

2 Name each part of the circle shown.

a 

b 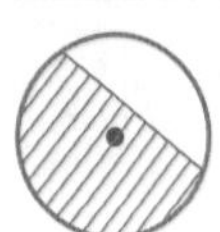

c 

d 

e 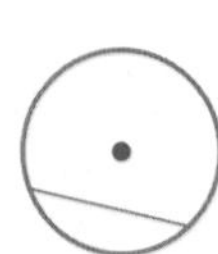

f 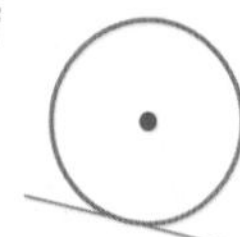

g 

h 

i 

j 

k 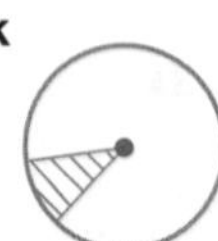

l 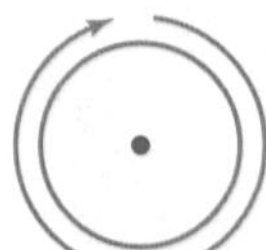

ALWAYS REMEMBER THAT A CIRCLE'S AREA IS MEASURED IN SQUARE UNITS, SO ITS FORMULA HAS r^2.

Teacher's tickbox

For each circle, calculate to 2 decimal places:

❏ its circumference ($C = \pi d$ or $C = 2\pi r$)   ❏ its area ($A = \pi r^2$)

1 2 cm

2 1.5 cm

3 8 cm

4 2 cm

5 5 cm

6 14 cm

7 0.5 cm

8 5 cm

9 6 cm

10 5.7 cm

11 10 cm

12 12 cm

13 9 cm

14 8.5 cm

15 7 cm

16 11 cm

17 3.8 cm

18 9.2 cm

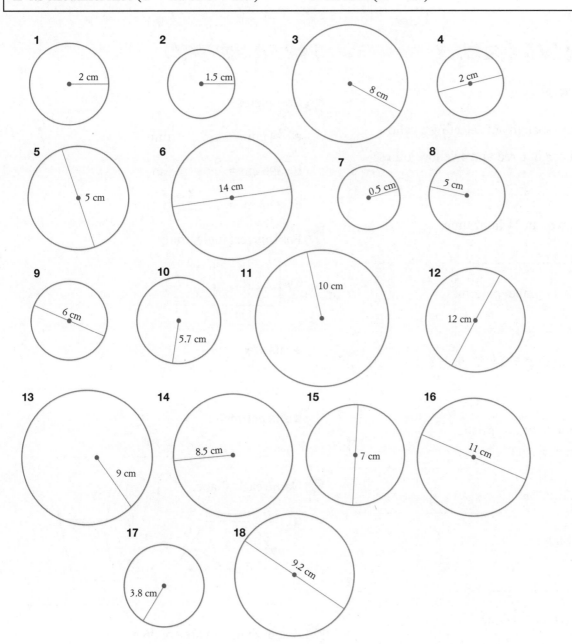

Mixed circumference answers (cm): 9.42, 43.98, 28.90, 18.85, 50.27, 34.56, 62.83, 35.81, 12.57, 23.88, 3.14, 31.42, 37.70, 53.41, 15.71, 56.55, 21.99, 6.28

Mixed area answers (cm^2): 254.47, 0.79, 113.10, 66.48, 12.57, 19.63, 38.48, 226.98, 78.54, 45.36, 153.94, 201.06, 3.14, 95.03, 102.07, 7.07, 314.16, 28.27

5 AREA 1

Name:

Due date:

Parent's signature:

Part A	/ 8 marks
Part B	/ 8 marks
Part C	/ 8 marks
Part D	/ 8 marks
Total	/ 32 marks

PART A: MENTAL MATHS

Calculators not allowed

1 What is the probability of selecting a yellow lolly from a bag of 6 red and 4 yellow lollies?

2 Write 10:08 p.m. in 24-hour time. ____________

3 Simplify 40 : 32. ____________

4 Complete this number pattern:

2, 4, 7, 11, ____________

5 Find the median of 7, 14, 11, 9, 3, 2.

6 Find the volume of this prism.

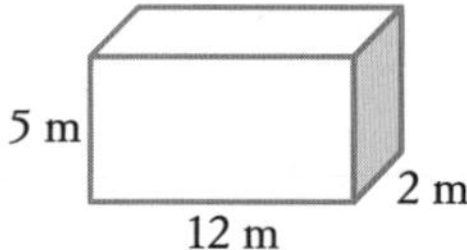

7 Find $\frac{3}{5}$ of \$400.

8 Evaluate \$162.60 − \$31.84.

PART B: REVIEW

1 Complete:

a 14 cm = ________ mm

b 390 cm = ________ m

c 72.5 m = ________ mm

2 For this rectangle, find:

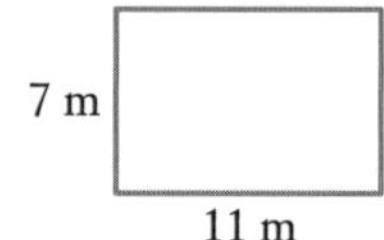

a its area

b its perimeter

3 Name each shape.

a 

b 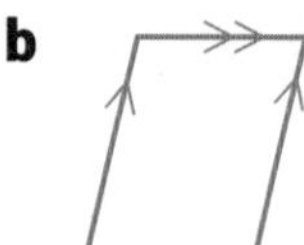

4 A square has an area of 36 m^2.

What is the length of each side?

9780170454520

PART C: PRACTICE

› Perimeter
› Metric units for area
› Areas of rectangles, triangles and parallelograms

1 Find the perimeter of each shape.

a

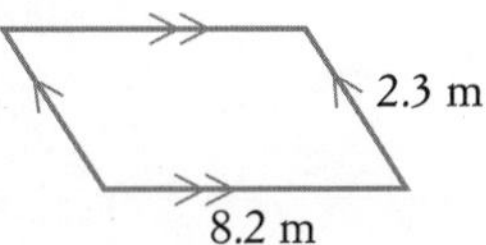

b

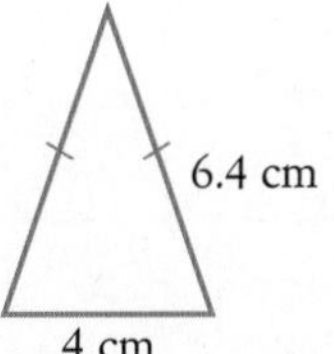

c

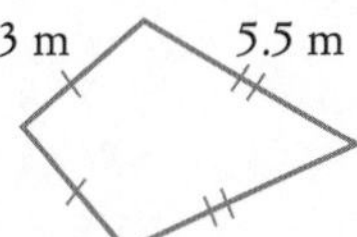

2 Complete:

a $0.7\text{ m}^2 =$ ________ cm^2

b $452\text{ mm}^2 =$ ________ cm^2

3 Find the area of each shape.

a

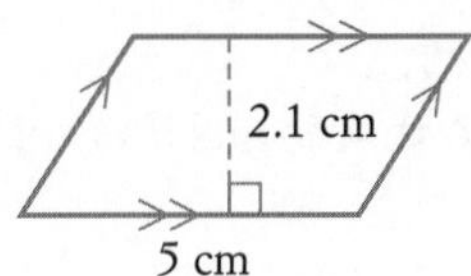

b

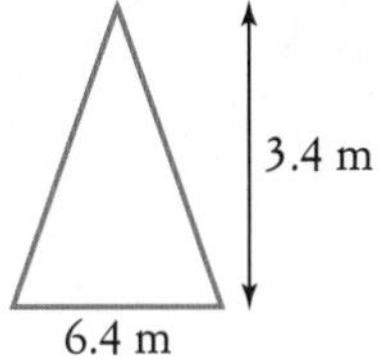

4 Write the formula for the area of a parallelogram.

PART D: NUMERACY AND LITERACY

1 What is the meaning of perimeter?

2 Write the metric unit of area 'cm²' in words.

3 Which unit of area is about the size of 2 soccer fields? ________

4 Find the perimeter of a rhombus with side lengths of 7.5 cm.

5 (2 marks) Explain the meaning of the formula $A = \frac{1}{2}bh$, what it is used for and what each variable stands for.

6 Find the area of this triangle in m^2.

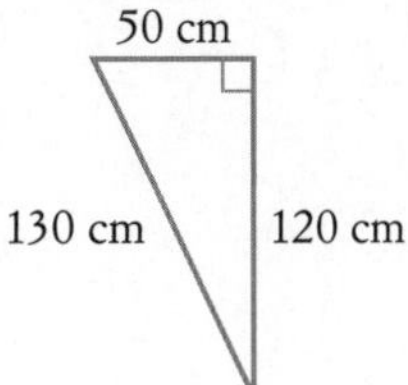

7 A parallelogram has an area of 54 cm^2. What is its perpendicular height if its base length is 4 cm?

HW HOMEWORK

C S F

Name:

Due date:

Parent's signature:

Part A	/ 8 marks
Part B	/ 8 marks
Part C	/ 8 marks
Part D	/ 8 marks
Total	/ 32 marks

LEARN THE FORMULAS AND METHODS FOR FINDING THE AREA OF A TRAPEZIUM, KITE AND RHOMBUS.

HW HOMEWORK

PART A: MENTAL MATHS

Calculators not allowed

1 Write 17:48 in 12-hour time. ______

2 Evaluate each expression.

a 90% of \$40. ______

b $220 + 180 \div (-9)$ ______

3 Complete this number pattern:

17, 13, 9, 5, ______

4 At the zoo excursion, there were 28 Year 7 students, 35 Year 8 students and 2 teachers. What is the probability of selecting at random from this group a Year 8 student?

______

5 Complete: 25 t = ______ kg.

6 Find the range of 7, 14, 11, 9, 3, 2.

______

7 Find the volume of this cube.

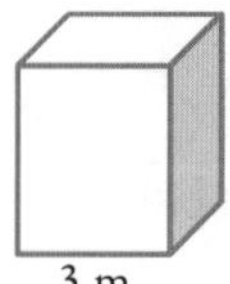

______

PART B: REVIEW

1 A square is 3.2 m long. Find:

a its area ______

b its perimeter. ______

2 Find the area of each shape in square metres.

a

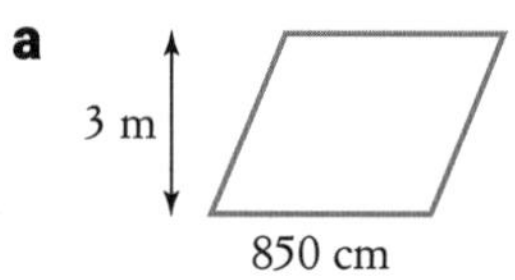

______

b

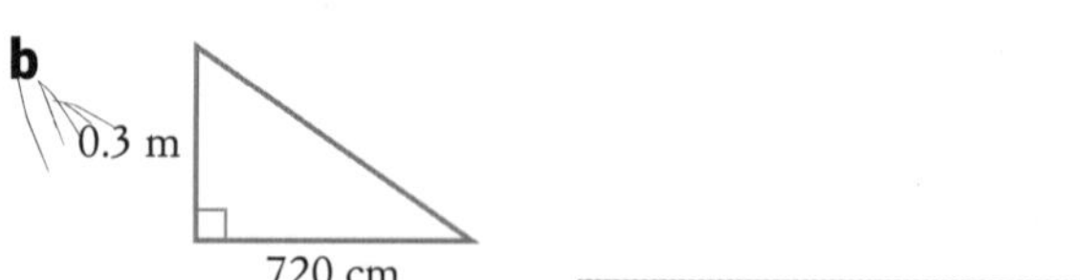

______

3 Complete:

a $9 \text{ m}^2 =$ ______ cm^2

b $1.3 \text{ km}^2 =$ ______ m^2

4 Complete this triangle so that it has an area of 15 m^2.

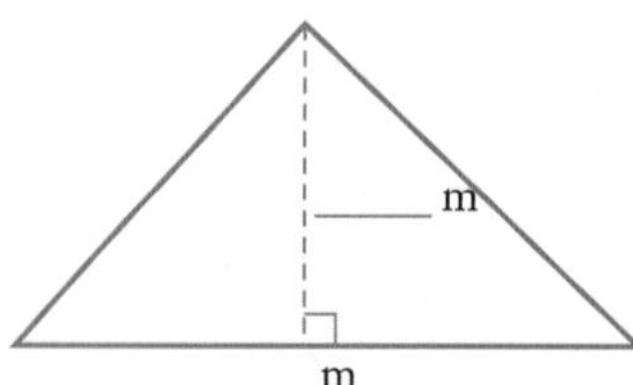

5 A rectangle has a length of 9 cm and a perimeter of 26 cm. What is its width?

______

C S F

9780170454520

PART C: PRACTICE

› Areas of composite shapes
› Area of a trapezium
› Areas of kites and rhombuses

1 For this shape, find:

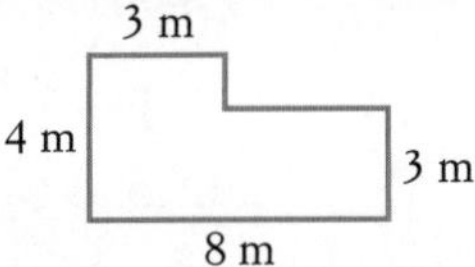

a its area ____________

b its perimeter. ____________

2 Find the area of each shape.

a

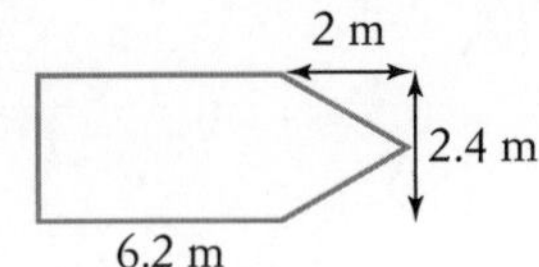

b

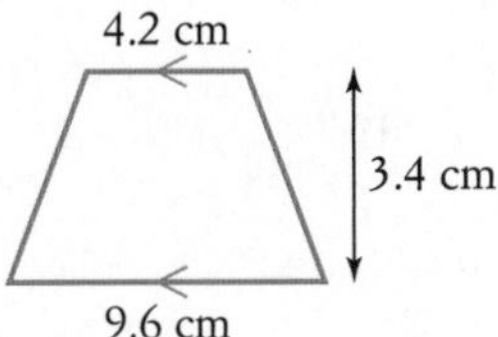

c

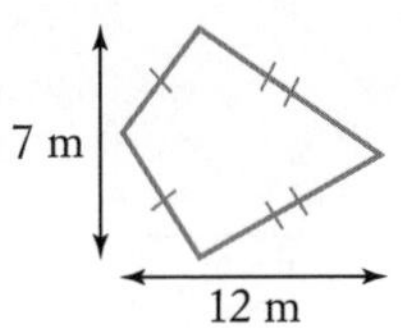

d

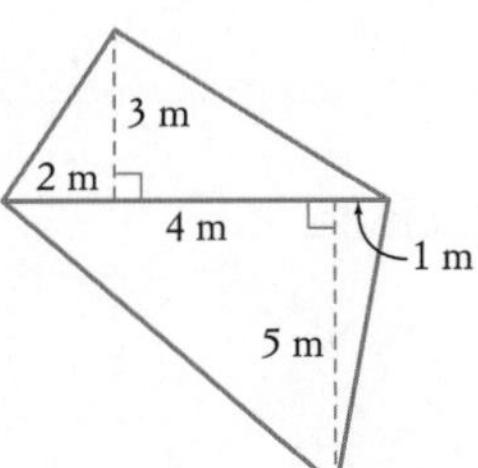

e

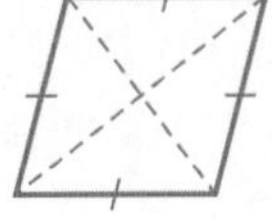

Diagonal 1 = 62 mm

Diagonal 2 = 45 mm

f

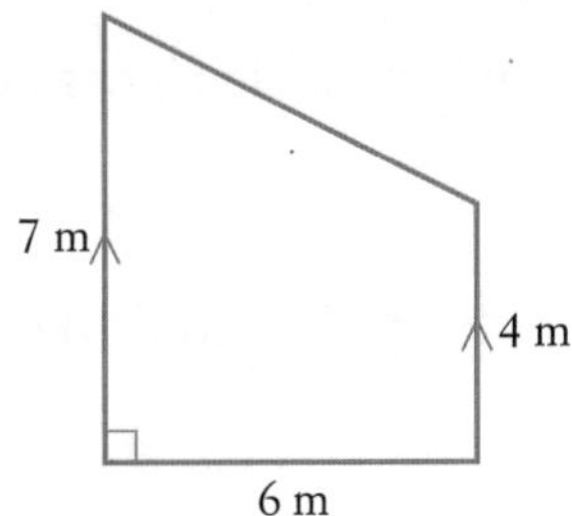

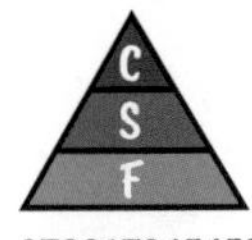

PART D: NUMERACY AND LITERACY

1 **a** What is the formula '$A = \frac{1}{2}(a + b)h$' used for?

____

b What does each variable in the formula stand for?

____

____

2 A kite has diagonals of length 65 cm and 86 cm. Find the area of the kite:

a in square centimetres

____

b in square metres.

____

3 A playground is an L-shape with dimensions south side 18 m, west side 8 m, upper north side 7 m, inner east side 4 m.

a Draw a diagram of the playground.

b Find its perimeter.

____

c Find its area.

____

____

4 Explain in words how to calculate the area of a rhombus.

____

____

HW HOMEWORK

5 CIRCUMFERENCE AND AREA OF A CIRCLE

WHAT ARE THE FORMULAS FOR THE CIRCUMFERENCE AND AREA OF A CIRCLE? BOTH INVOLVE PI, BUT IS R SQUARED OR NOT SQUARED?

Name:

Due date:

Parent's signature:

Part A	/ 8 marks
Part B	/ 8 marks
Part C	/ 8 marks
Part D	/ 8 marks
Total	/ 32 marks

PART A: MENTAL MATHS

Calculators not allowed

1 Evaluate $(-5)^3$ ______

2 Simplify $5mn - 3mn + 9$. ______

3 If a letter from the word MATHEMATICS is chosen at random, what is the probability that it is not a vowel? ______

4 Convert 340 minutes to hours and minutes.

______

5 Convert $\frac{2}{3}$ to a decimal. ______

6 Classify this triangle:

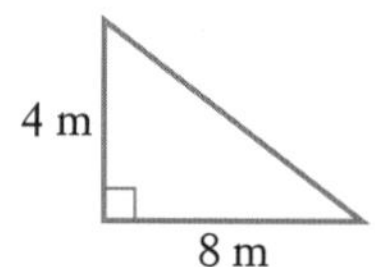

a by sides ______

b by angles. ______

7 Find correct to one decimal place the mean of 7, 14, 11, 9, 3, 2.

______

PART B: REVIEW

1 Find the area of the triangle in Question **6** of Part A.

______

2 Complete:

a 5.6 ha = ______ m^2

b 46 850 cm^2 = ______ m^2

3 Find the area of each shape.

a

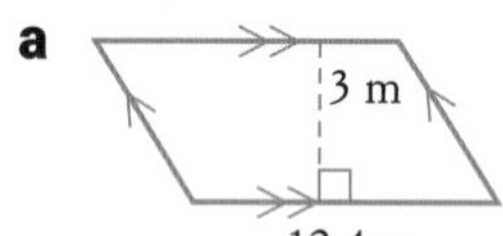

b

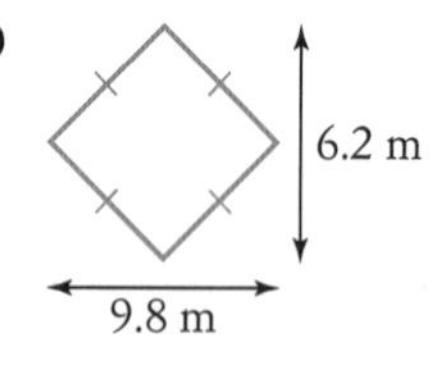

______ ______

______ ______

4 Draw the **radius** of this circle.

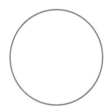

5 Find the perimeter of:

a an equilateral triangle with side length 9 cm

______

b a kite with sides of length 3 m and 10 m

______

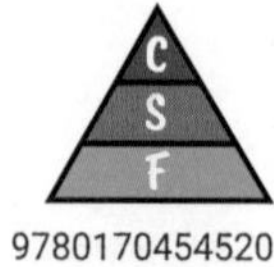

9780170454520

PART C: PRACTICE

› Circumference of a circle
› Area of a circle

For this page, express answers to 2 decimal places if needed.

1 Find the perimeter of each shape.

a

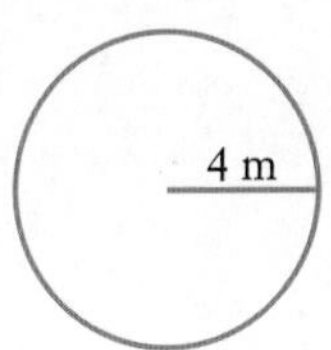

b

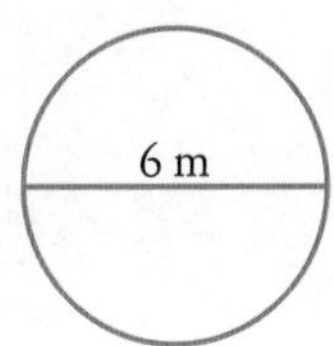

c

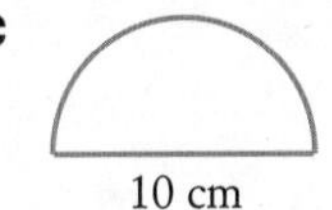

d

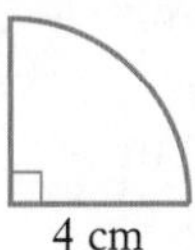

2 Find the area of each shape.

a

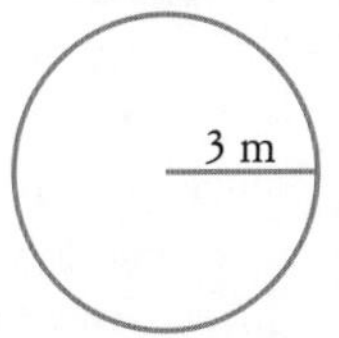

b

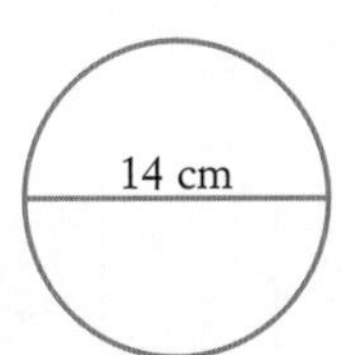

c

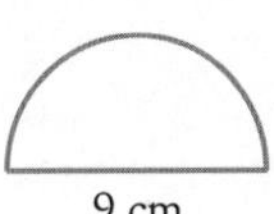

d

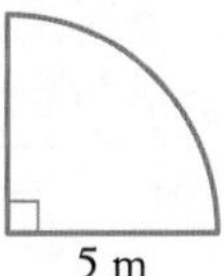
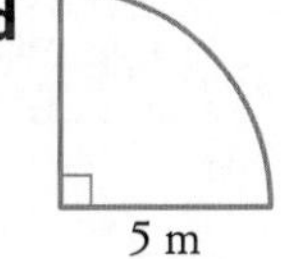

PART D: NUMERACY AND LITERACY

1 What type of number is π?

2 What does **circumference** mean?

3 **a** Draw a semi-circle with diameter 6.5 cm.

b Find its area.

c How many axes of symmetry has a semi-circle?

4 Write down the formula for the area of a circle.

5 **a** Find the circumference of a circular tablecloth if its diameter is 3 m.

b How many square metres of material would you need to make this tablecloth?

HW HOMEWORK

C S F

5 VOLUME AND CAPACITY

VOLUME IS HOW MUCH SPACE A 3D (SOLID) SHAPE TAKES UP, MEASURED IN CUBIC UNITS.

Name:

Due date:

Parent's signature:

Part A	/ 8 marks
Part B	/ 8 marks
Part C	/ 8 marks
Part D	/ 8 marks
Total	/ 32 marks

HW HOMEWORK

PART A: MENTAL MATHS

Calculators not allowed

1 From a bag with 7 red, 4 blue and 9 black marbles, find the percentage probability of randomly selecting a marble that is not blue. ______

1 What is the time 4 hours and 20 minutes after 13:45? ______

3 Evaluate each expression.

a $\frac{3}{5} - \frac{1}{3}$

______

b $5.24 \div 0.4$ ______

4 Simplify $9 \times a \times (-2) \times d$. ______

5 Find the median of these data values.

Value	3	4	5	6	7	8
Number of dots	2	3	3	1	0	1

6 **a** Sketch a regular pentagon.

b Find the perimeter of a regular pentagon with a side length of 4 m.

______

PART B: REVIEW

1 Complete:

a $1.2\ \text{m}^2 =$ ______ cm^2

b $39\ 000\ \text{mm}^2 =$ ______ cm^2

c $100\ 000\ \text{m}^2 =$ ______ ha

2 Find the area of each shape.

a

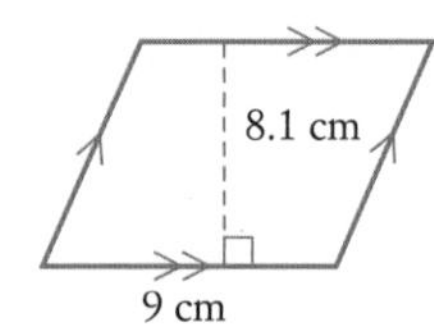

______

b

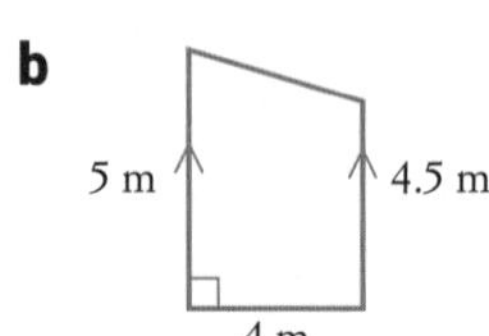

______

c

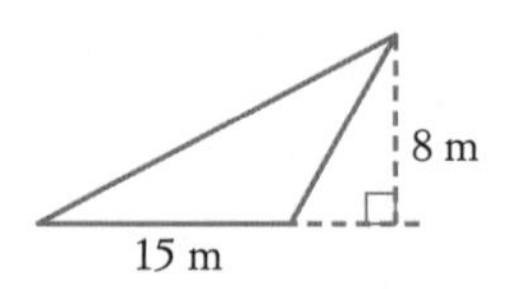

______

3 Find the volume of this cube.

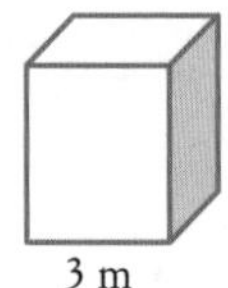

______

C S F

9780170454520

4 Find correct to 2 decimal places the area of this circle.

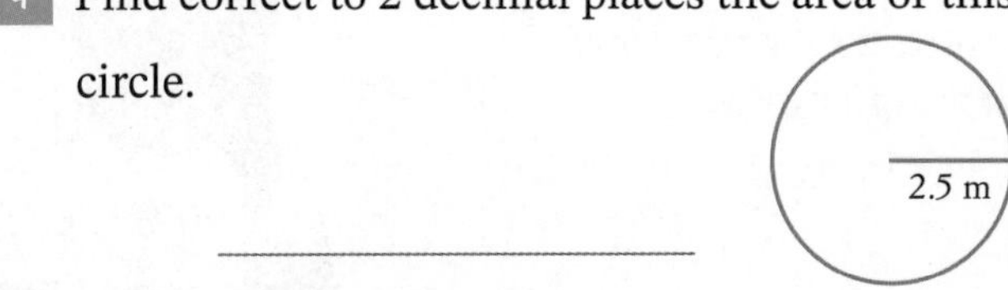

PART C: PRACTICE

› Metric units for volume
› Volumes of prisms and cylinders
› Volume and capacity

1 Complete:

a $3.2\text{ m}^3 =$ ________ cm^3

b $108\,000\text{ mm}^3 =$ ________ cm^3

2 Find the volume of each prism.

a

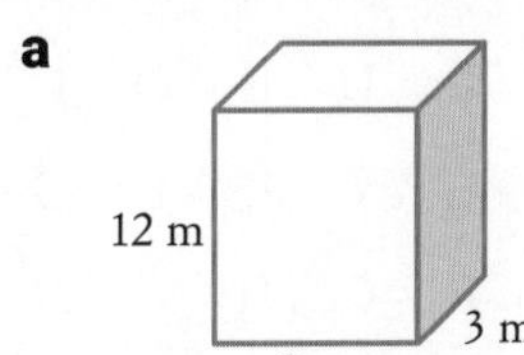

b

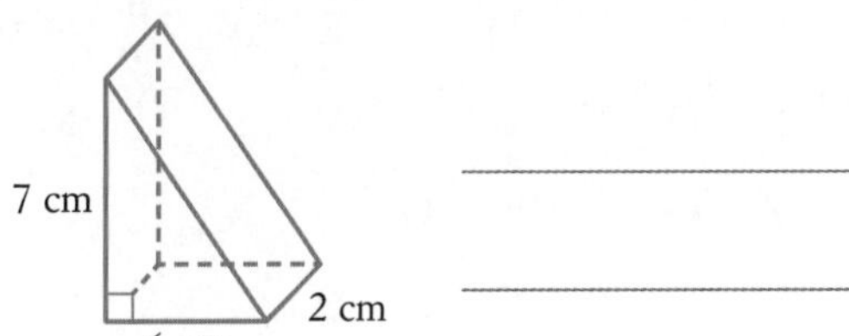

c

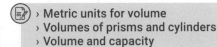

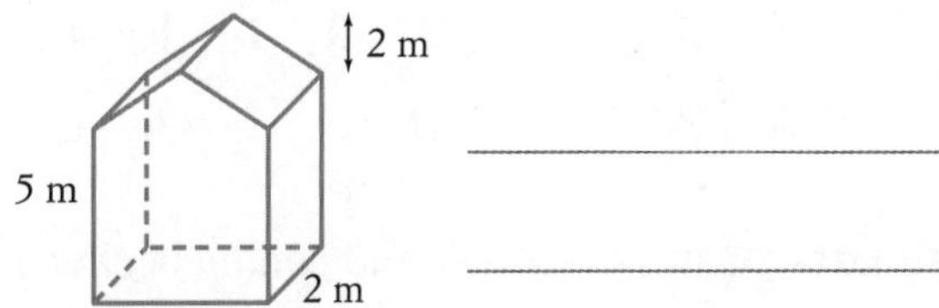

3 **a** What shape is a cross-section of this cylinder?

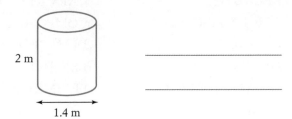

2 m

1.4 m

b Find correct to 2 decimal places the volume of the cylinder. ________

c Find the capacity of this cylinder to the nearest litre.

PART D: NUMERACY AND LITERACY

1 What is the meaning of **volume**?

2 Draw a square prism.

3 How many litres of water can a cube of length 1 m hold? ________

4 **a** A drink container in the shape of a prism has a base area of 40 cm^2 and a height of 7 cm. Find its volume.

b Would this container hold 300 mL of apple juice? ________

5 **a** What is the formula for the volume of a cylinder?

b What does each variable in the formula stand for?

6 A can of soup has a base radius of 5 cm and a height of 9 cm. Find the capacity of this can correct to the nearest mL.

PUZZLE SHEET

5 AREA AND VOLUME FIND-A-WORD

THIS PUZZLE INCLUDES THE IMPORTANT KEYWORDS FROM THIS TOPIC.

U V M B F H E I G H T R J S V H E L F L D P K W C
N E T H D V D I T L A U Z L L X C J W H R Q A W R
Y T P A R A L L E L O G R A M V Y N G E G R R N O
A M V Q W W K H F O R M U L A R T O T F E A E Z S
R Z T C O M P O S I T E S O L I D E U A H M C W S
E O T V S K B A Q U A D R A N T M T B W N E T H S
W U T S N X V D B A S E S F N A N C I G O D A O E
W H B C M F H R Q S X F S K I Z V E C I I I N N C
O W A F E R A D I U S C B D M K D S B U D S G K T
R F K N K S T R I A N G L E P L N S L W L U L O I
P C A P A C I T Y V Q Z C G C L F O V C U V E F O
J S M J R A L U G N A I R T E U A R P D R O H C N
G Z G P E R I M E T E R S N U O T C X I R R G O S
L Q M R M U I Z E P A R T C Y L I N D E R A V D U
F I H C Y H E C T A R E E M U L O V L A V S U R U
E C N E R E F M U C R I C W T K Q H T G N E L Q C
L P H A P G U U I Y I E B E X C I R C L E I A C S
L M N R A L U C I D N E P R E P O E C A F R U S Z
B R E A D T H A U K R E C T A N G U L A R I R J X
M S I R P Q V T N E M E R U S A E M G D B W A O X

Find these words in the puzzle above. They are across, up and down, and diagonal, and can be backwards as well as forwards.

AREA	BASE	BREADTH	CAPACITY
CHORD	CIRCLE	CIRCUMFERENCE	COMPOSITE
CROSS-SECTION	CUBIC	CYLINDER	DIAMETER
FORMULA	HECTARE	HEIGHT	LENGTH
MEASUREMENT	NET	PARALLELOGRAM	PERIMETER
PERPENDICULAR	PI	PRISM	QUADRANT
RADIUS	RECTANGLE	RECTANGULAR	SECTOR
SIDE	SOLID	SQUARE	SURFACE
TRAPEZIUM	TRIANGLE	TRIANGULAR	VOLUME

THIS ASSIGNMENT WILL HELP YOU PREPARE FOR THE FRACTIONS AND PERCENTAGES TOPIC. TRY DOING THESE QUESTIONS WITHOUT A CALCULATOR.

WS WORKSHEET

PART A: BASIC SKILLS

/ 15 marks

1 What angle is supplementary to 35°?

____

2 Evaluate $\sqrt[3]{729}$. ______________________

3 What type of triangle has no equal sides?

____

4 How many hours and minutes between 7:45 a.m. and 1:35 p.m.? ______________________

5 Write an algebraic expression for '5 more than double *x*'. ______________________

6 Simplify $\frac{24}{90}$. ______________________

7 In which quadrant of the number plane is the point $(-3, -1)$? ______________________

8 Calculate the area of this figure.

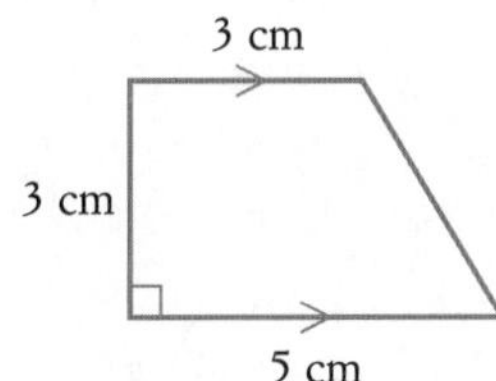

9 Expand $-7(2a + 1)$. ______________________

10 Complete: Any number that is divisible by 5 ends with the digit __________ or __________.

11 If $r = 3$, then evaluate $2r + 5$. ______________________

12 Evaluate $2 \times (-4) + (-18) \div 9$.

____

13 Simplify $3y + 5x - 7x + 2y$.

____

14 What order of rotational symmetry has a square? ______________________

15 Test whether this triangle is right-angled.

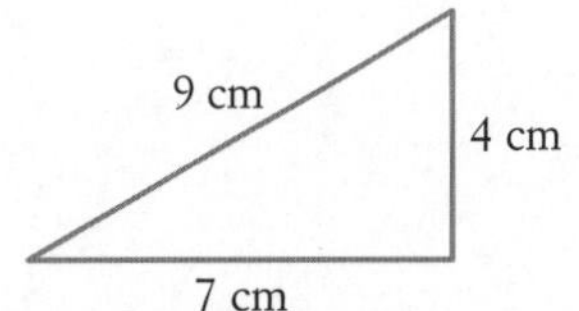

PART B: NUMBER

/ 25 marks

16 Write these decimals in ascending order:

0.3, 0.317, 0.31, 0.37, 0.371.

____

17 List the 9 factors of 100.

____

18 Simplify:

a $\frac{18}{100}$ ______________________

b $\frac{72}{100}$ ______________________

c $\frac{35}{100}$ ______________________

19 Evaluate:

a $8.5 \div 100$ ______________________

b 0.9×100 ______________________

c $\frac{3}{5} \times 100$ ______________________

d $0.13 \times \$210$ ______________________

WS WORKSHEET

20 What fraction is the same as 50%?

____

21 Find 10% of $450. ______________________

22 Decrease 56 by 11. ______________________

23 What is $\frac{3}{4}$ as a percentage?

____

24 Convert each fraction to a decimal.

a $\frac{27}{50}$ ______________________

b $\frac{3}{8}$ ______________________

c $\frac{7}{12}$ ______________________

25 Peter bought a computer for $2150 and sold it for $1670. How much did he lose?

____

26 How many:

a minutes in 2 hours? ______________________

b mL in 5.4 L? ______________________

c cm in 2.7 m? ______________________

27 Convert each decimal to a simple fraction.

a 0.85 ______________________

b 0.24 ______________________

c 0.4 ______________________

28 Increase 14 by 32. ______________________

29 If 46% of the people at a meeting were male, what percentage were female?

____

PART C: CHALLENGE

Bonus / 3 marks

2 adults and 2 children need to cross a river but their canoe can only carry one adult or 2 children at a time. Determine how they can cross the river in 9 trips.

C
S
F

9780170454520

FRACTIONS AND PERCENTAGES FIND-A-WORD 

HERE ARE THE LETTERS, THERE ARE THE WORDS. YOU KNOW WHAT TO DO!

```
K I W L K R I Y I C S O D N U R J O J Y J X P S A
Q H K W D A Z J X A O E Y N K P P B O B S O U K L
I L D A R K T C W C T M I V S P G F E P V T K I C
G E E T O U Z G T K L T M I R R T Y G D W B R X R
W S R D T E P M Y B A A M I E F Z I X I D I A Z X
E A D H A T E T A R P P S D S M H N F S V T M H N
W E N H N A R A Y Z L R U C O S W A S C C N R D I
O R U K I L C A P I F C O S E F I O Q O K E H E F
S C H F M U E W F P T G G P P N A O N U C L S S M
Y N V A O C N Y X I I P B S E G D V N N G A X C W
R I G X N L T X O T C M N F T R E I V T R V T E L
E F G Z E A A N I L N O P T D R O R N P X I I N Q
D T N Q D C G W S Y M F J R T E P S R G I U F D U
R C U U V Q E Q R M T I R I O S C I S B Z Q O I E
O F T D Q K W N O E T I K A A P C I G O E E R N L
Z U Q T P J E C Y C V Z T L C E E G M W L D P G C
T S E R E T N I T H O V E N I T B R T A C V A C D
Q E S A E R C E D W O S E Z A D I G N I L L E S V
C B L C E Z O B R V M V T H L U Q O Y P F Q P W R
V S U R O T A R E M U N A I L N Q J N O X G Z C C
```

Find these words in the puzzle above. They are across, up and down, and diagonal, and can be backwards as well as forwards.

ASCENDING	CALCULATE	COMMISSION	COMMON
CONVERT	COST	DECIMAL	DECREASE
DENOMINATOR	DESCENDING	DISCOUNT	EQUIVALENT
FRACTION	GST	HUNDRED	IMPROPER
INCREASE	INTEREST	LOSS	MARK-UP
NUMERATOR	ORDER	PERCENTAGE	PRICE
PROPER	PROFIT	QUANTITY	REDUCTION
SALE	SELLING	SIMPLIFY	UNITARY

6 PERCENTAGE SHORTCUTS

WS WORKSHEET

ONE HANDY MATHS SKILL IS TO BE ABLE TO CONVERT A PERCENTAGE TO A FRACTION IN YOUR HEAD. IT WILL SAVE YOU A LOT OF TIME IN THE FUTURE.

When solving percentage problems, often it is more convenient to type percentages as decimals into the calculator.

1 Write each percentage as a decimal.

- **a** 12%
- **b** 73%
- **c** 5%
- **d** 40%
- **e** 18.6%
- **f** 8%
- **g** 3.1%
- **h** 122%
- **i** 6.95%
- **j** $12\frac{1}{2}\%$
- **k** 150%
- **l** $8\frac{1}{4}\%$

Example *12% of \$91 = 0.12 × 91 = \$10.92*

2 Evaluate each expression.

- **a** 24% of \$60
- **b** 81% of \$491
- **c** 7% of \$36
- **d** 20% of \$77.30
- **e** 44.5% of \$320
- **f** 8.9% of \$134 000
- **g** 1% of \$74
- **h** 9.25% of \$49
- **i** 18% of \$296
- **j** $6\frac{1}{2}\%$ of \$2000
- **k** 118% of \$54
- **l** 16.3% of \$23.50

Example *Increase \$64 by 15%*

100% + 15% = 115%

115% × \$64 = 1.15 × 64 = \$73.60

(Check this gives the correct answer!)

3 Increase:

- **a** \$79 by 15%
- **b** \$30 by 25%
- **c** \$128 by 40%
- **d** \$340 by 6%
- **e** \$22.30 by 11%
- **f** \$395 by 5%
- **g** \$75.40 by 10%
- **h** \$220 by 16%
- **i** \$381 by $7\frac{1}{4}\%$
- **j** \$42.20 by 12.8%
- **k** \$528.60 by 8.3%
- **l** \$45.50 by 100%

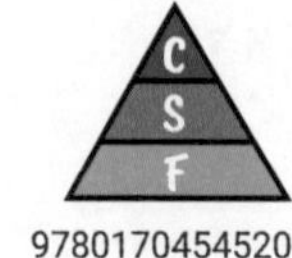

Example	*Decrease (discount) \$45 by 10%*
	100% − 10% = 90%
	90% × \$45 = 0.9 × 45 = \$40.50

4 Decrease:

- **a** \$30 by 10%
- **b** \$75 by 8%
- **c** \$800 by 5%
- **d** \$179 by 15%
- **e** \$88.50 by 12%
- **f** \$460 by 18%
- **g** \$7440 by 30%
- **h** \$1050 by 45%
- **i** \$79.90 by $9\frac{3}{4}$%
- **j** \$235 by 5.2%
- **k** \$67 by 66%
- **l** \$380.10 by 33%

5 Georgia earns 5% commission on all her sales of kitchenware. How much will she earn from selling \$8750 worth of kitchenware?

____

6 The Jean Pool is having a '12% off' sale. Calculate the sale price of each item.

- **a** jeans \$74.60

  ____

- **b** caps \$12.80

  ____

- **c** shirts \$37.50

  ____

7 Patrick earns a salary of \$78 290 but received a pay rise of 8.5%. Calculate his new salary.

____

8 A sum of \$5000 is invested in a savings account that earns 7% interest each year.

- **a** Complete: Increasing \$5000 by 7% is the same as multiplying \$5000 by ____________

  ____

- **b** Calculate the size of the \$5000 investment after 1 year.

  ____

- **c** Increase the answer in part **b** by 7% to find the size of the investment after another year.

  ____

- **d** How much is the investment worth after another year (over 3 years)?

  ____

- **e** **Challenge:** What is the meaning of this formula: $A = 5000 \times (1.07)^n$?

  ____

WS WORKSHEET

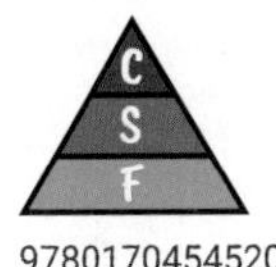

6 PERCENTAGES WITHOUT CALCULATORS

THAT'S RIGHT ... YOU CAN DO THESE IN YOUR HEAD OR USING PEN AND PAPER.

WS WORKSHEET

Calculators not allowed

1 Convert to a simple fraction:

a 0.8 ____________

b 0.25 ____________

c 0.015 ____________

d 0.52 ____________

2 Convert to a decimal:

a $\frac{14}{100}$ ____________

b $\frac{1}{8}$ ____________

c $\frac{2}{3}$ ____________

d $\frac{3}{5}$ ____________

3 Evaluate:

a $\frac{1}{6} \times 48$ ____________

b $\frac{2}{5} \times 35$ ____________

c $\frac{2}{3} \times 27$ ____________

d $\frac{5}{8} \times 40$ ____________

e $\frac{1}{2} \times \$154$ ____________

f $\frac{7}{10}$ of 1 hour ____________

g $\frac{3}{4}$ of 1 kg ____________

h $\frac{4}{5}$ of 1 L ____________

4 Convert to a simple fraction

a 2% ____________

b $66\frac{2}{3}\%$ ____________

c 44% ____________

d 85% ____________

5 Convert to a decimal:

a 78% ____________

b 8% ____________

c $56\frac{1}{4}\%$ ____________

d 32.09% ____________

6 Evaluate:

a $28.75 \div 100$ ____________

b $1.62 \div 10$ ____________

c $3404 \div 1000$ ____________

d 0.57×100 ____________

e 30×80 ____________

f 25×20 ____________

7 Convert to a percentage:

a $\frac{7}{10}$ ____________

b $\frac{1}{5}$ ____________

c $\frac{1}{3}$ ____________

d $\frac{3}{4}$ ____________

e 0.29 ____________

f 0.9 ____________

g 0.091 ____________

h 0.475 ____________

C S F

9780170454520

8 Evaluate:

a 0.25×20 ______

b 0.7×40 ______

c 0.2 of 1 minute ______

d 0.5 of 1 year ______

e $25\% \times 28$ ______

f 20% of 45 ______

g 5% of $120 ______

h 8% of $70 ______

i $33\frac{1}{3}\%$ of 1 day ______

j 75% of 1 hour ______

k 10% of $35.40 ______

l 60% of 2 metres ______

9 Express as a simple fraction:

a 50 cm of 3 metres ______

b 42 minutes of 1 hour ______

c 40c of $2 ______

d 120 g of 1 kg ______

10 Write in descending order: $\frac{3}{4}$, 0.78, 0.8, 87%

______

11 If $\frac{2}{3}$ of a number is 18, what is the number?

______

12 If 40% of a number is 56, what is the number?

______

13 Increase:

a $60 by $33\frac{1}{3}\%$ ______

b $36 by 25% ______

c $28 by 10% ______

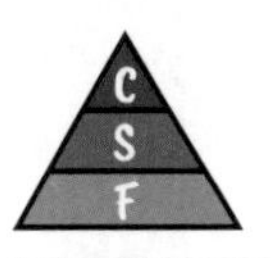

14 Express as a percentage:

a 15 out of 25 ______

b $66 of $200 ______

c 9 months of 1 year ______

d 125 mL of 1 L ______

15 Evaluate:

a 0.4×0.3 ______

b 8×0.8 ______

c 0.2×7 ______

d 0.6×0.5 ______

16 Decrease:

a $50 by 40% ______

b $300 by 2% ______

c $140 by 15% ______

17 If 15% of a number is 75, what is the number?

______

18 Gina scored 32 out of 40 in a test. What was her percentage mark? ______

19 Harley bought a motorbike for $25 000, then later sold it at a 20% loss. What was his selling price?

20 It cost Melanie $60 to make a set of drawers. For how much would she need to sell them to make a profit of 15%? ______

21 If $\frac{7}{10}$ of a number is 56, what is the number?

______

22 Paul weighed 96 kg but lost 12 kg after a strict diet. What percentage of his weight did Paul lose?

______

23 Find the percentage discount on a jacket whose price has been reduced from $80 to $64.

______

6 FRACTIONS

A LOT OF STUDENTS HAVE TROUBLE WITH FRACTIONS, SO IT HELPS IF YOU PRACTISE UNTIL YOU GET IT RIGHT. ASK FOR HELP IF YOU NEED.

Name:

Due date:

Parent's signature:

Part A	/ 8 marks
Part B	/ 8 marks
Part C	/ 8 marks
Part D	/ 8 marks
Total	/ 32 marks

PART A: MENTAL MATHS

Calculators not allowed

1 Evaluate 4^3. ______

2 Complete: An isosceles triangle is a triangle with

______

3 Write a number than can be rounded to 3.55.

______

4 Find the range of these values.

5 Simplify each expression:

a $\dfrac{45a}{-9}$ ______

b $4k - 6 + 6 - 3k$ ______

6 What is an octagon?

______

______

7 Write an algebraic expression for the number 5 less than the sum of m and n.

______

PART B: REVIEW

1 Simplify each fraction.

a $\dfrac{10}{12}$ ______

b $\dfrac{18}{45}$ ______

2 Convert each fraction to a mixed numeral.

a $\dfrac{23}{7}$ ______

b $\dfrac{16}{12}$ ______

3 Which fraction is smaller: $\dfrac{8}{9}$ or $\dfrac{5}{6}$?

______

______

4 Complete: $\dfrac{24}{30} = \dfrac{}{10}$.

5 Convert $4\dfrac{3}{5}$ to an improper fraction.

______

______

6 Evaluate $\dfrac{11}{12} - \dfrac{5}{12}$.

______

______

HW HOMEWORK

9780170454520

PART C: PRACTICE

› Adding and subtracting fractions
› Multiplying and dividing fractions

1 Evaluate each expression.

a $\frac{4}{5}+\frac{2}{3}$

b $\frac{7}{8}-\frac{3}{4}$

c $1\frac{4}{5}+2\frac{1}{3}$

d $4-2\frac{2}{3}$

e $\frac{2}{3}\times\frac{6}{5}$

f $\frac{4}{9}\div\frac{8}{27}$

g $1\frac{3}{4}\times4\frac{3}{5}$

h $5\frac{1}{4}\div3$

C S F

PART D: NUMERACY AND LITERACY

1 **a** What is the **reciprocal** of $\frac{3}{8}$, as a mixed numeral?

b What happens when you multiply a fraction by its reciprocal?

2 Three friends had a box of chocolates. Jordan ate $\frac{3}{8}$ of the chocolates, Kieran ate $\frac{3}{10}$ and Sammy ate $\frac{1}{4}$.

a What fraction of the chocolates were left?

b If Jordan ate 15 chocolates, how many chocolates were in the box originally?

3 Complete:

a To add or subtract fractions, we should first convert them to equivalent fractions with the same ______________.

b To divide by a fraction, ______________ by its reciprocal.

4 Evaluate each expression.

a $\left(\frac{2}{3}\right)^2$ ______________

b $\sqrt{\frac{25}{64}}$ ______________

HW HOMEWORK

6 PERCENTAGES 1

PERCENTAGES ARE FOUND EVERYWHERE AND EVERY DAY. LEARN HOW TO USE THEM, AND YOU'LL GET BETTER AT MATHS.

Name:

Due date:

Parent's signature:

Part A	/ 8 marks
Part B	/ 8 marks
Part C	/ 8 marks
Part D	/ 8 marks
Total	/ 32 marks

HW HOMEWORK

PART A: MENTAL MATHS

Calculators not allowed

1 Evaluate each expression.

a $\sqrt[3]{125}$

______

b $\sqrt{10^2 - 6^2}$

______

2 Find the mode of this data set.

3 Write an algebraic expression for the number of minutes in x hours.

______

4 An equilateral triangle has a perimeter of 22.5 cm. What is the length of one side?

______

5 Simplify $n \times n \times 1 \times n$. ______

6 What is the angle sum of a quadrilateral?

______

7 If one angle in a parallelogram is 75°, what are the sizes of the other 3 angles?

______

PART B: REVIEW

1 Convert each percentage to a simple fraction.

a 28%

______

b 55%

______

2 Convert each percentage to a decimal.

a 7% ______

b 18.5% ______

3 Find $\frac{2}{3}$ of 4 hours (in hours and minutes)

______

4 Arrange these numbers in ascending order.

$0.4 \quad \frac{3}{4} \quad 65\% \quad \frac{7}{8}$

______

______

______

5 Convert each number to a percentage.

a $\frac{22}{25}$

______

b 0.64

______

C S F

9780170454520

PART C: PRACTICE

› Percentage of a quantity
› Expressing amounts as fractions and percentages
› Percentage increase and decrease

1 Find:

a 70% of $550

b 32% of 3 L (in mL)

2 Express 45 minutes out of 2 hours:

a as a simple fraction ____________

b as a percentage ____________

3 Yasmin's weekly pay of $920 increased by 6%. What is her new pay?

4 Express each as a percentage.

a $5.88 out of $84

b 175 cm out of 5 m

5 An electronics store is having a '15% off' sale. Find the sale price of an eReader with an original price of $255.

PART D: NUMERACY AND LITERACY

1 Explain how to convert:

a $\frac{2}{3}$ to a percentage

b 70% to a simple fraction

2 Complete:

a When calculating a percentage increase, we ____________ the new amount to the original.

b Increasing an amount by 40% is the same as finding ____________ % of the amount.

c Decreasing an amount by 40% is the same as finding ____________ % of the amount.

3 (2 marks) Explain how to find 45 mm as a percentage of 5 cm.

4 Mitchell needs to score more than 85% to pass his driving knowledge test. At least how many questions must he answer correctly if there are 25 questions in the test?

HW HOMEWORK

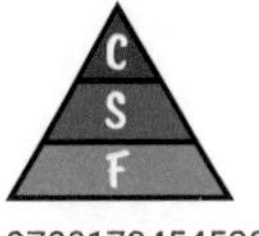

6 PERCENTAGES 2

THIS ASSIGNMENT LOOKS
AT PERCENTAGE PROBLEMS.
COMPLETE THE QUESTIONS,
AND YOU'LL BE A % PRO!

Name:	
Due date:	
Parent's signature:	

Part A	/ 8 marks
Part B	/ 8 marks
Part C	/ 8 marks
Part D	/ 8 marks
Total	/ 32 marks

PART A: MENTAL MATHS

Calculators not allowed

1 Evaluate each expression.

a $\sqrt{4\times4\times6\times6}$ ____________

b 0.9×0.06 ____________

2 Solve the equation $\frac{w+4}{3} = -8$.

____

3 **a** Find the outlier of this set of data.

b Find the mean of the scores above.

____

4 A triangle has a base 8 m long and an area of 20 m^2. What is its perpendicular height?

____

5 **a** Mark a pair of alternate angles.

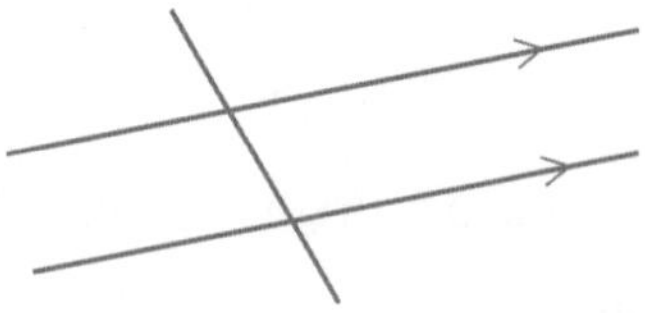

b What is the property of alternate angles on parallel lines?

____

PART B: REVIEW

Calculators not allowed

1 Convert $\frac{3}{4}$ to a percentage. ____________

2 Find $\frac{3}{8}$ of \$32.

____

3 12% of a school's students are left-handed. How many is this if the student population is 800?

____

4 Convert each percentage to a simple fraction.

a 40%

____

b $12\frac{1}{2}\%$

____

5 Asabi scored 36 out of 40 in her maths test. Calculate this mark as a percentage.

____

6 Find $33\frac{1}{3}\%$ of \$75.

____

7 Calculate the sale price of a home gym marked at \$480 after a 10% discount.

____

C S F

HW HOMEWORK

9780170454520

PART C: PRACTICE

› The unitary method
› Profit and loss
› Percentage problems

1 A restaurant bill was $148. Calculate the final amount after 10% GST is added.

____

2 A shop bought skateboards for $180 and sold them for $215.

a What is the cost price? ______________

b How much is the profit?

____

c Calculate, correct to one decimal place, the profit as a percentage of the cost price.

____

3 Alice saves 12% of her weekly pay for a holiday. If she saves $72, what is her weekly pay?

____

____

4 Tim bought a car for $16 000 and sold it for $13 800.

a How much is the loss? ______________

b Calculate the loss as a percentage of the cost price.

____

5 A taxi fare was $68.75 after 10% GST was added. Calculate the cost before GST.

____

____

PART D: NUMERACY AND LITERACY

1 What does GST stand for?

____

2 Miley earns 2.5% commission for selling a $423 000 home. How much does she earn?

____

3 What is the name given to the difference between the cost price and selling price if:

a the cost price is higher?

____

b the selling price is higher?

____

4 Lina's team played 40 matches this season and won 31 of them. What percentage of matches did they win?

____

5 Complete:

a To find a percentage profit of the cost price, divide the profit by the ______________ and multiply by ______________.

b To convert a percentage to a fraction, first write it with a ______________ of ______________.

6 Ashok earned a 8% bonus on his salary for completing a project on time. If the bonus was $6608, what is his salary?

____

____

HW HOMEWORK

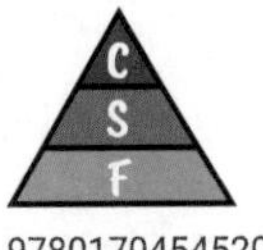

6 FRACTIONS AND PERCENTAGES REVISION

WE'RE UP TO THE LAST ASSIGNMENT FOR THIS TOPIC. HAVE YOU MASTERED FRACTIONS AND PERCENTAGES NOW?

Name:

Due date:

Parent's signature:

Part A	/ 8 marks
Part B	/ 8 marks
Part C	/ 8 marks
Part D	/ 8 marks
Total	/ 32 marks

PART A: MENTAL MATHS

Calculators not allowed

1 How many axes of symmetry has a square?

2 Complete: 4.2 kL = ______ L

3 Evaluate each expression.

a $\sqrt{3\times3\times7\times7}$

b $2.91 \div 0.6$

c $16 \div (-4)^2$

4 A letter is selected at random from the alphabet. What is the probability that it is a vowel or Y?

5 Simplify $11m \times (-4n)$.

6 Find x.

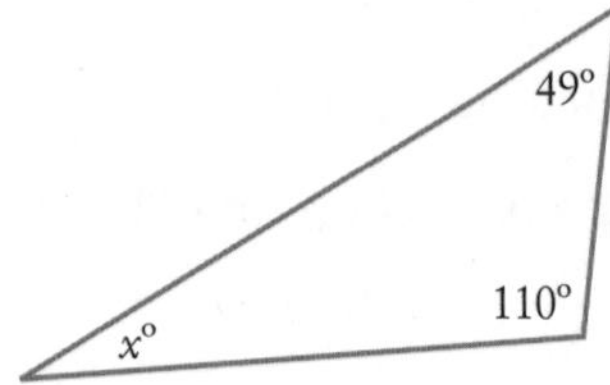

PART B: REVIEW

1 Simplify $\frac{32}{12}$.

2 Arrange these numbers in ascending order.

$$75\%, \frac{4}{5}, \frac{7}{8}, 0.\dot{7}$$

3 Convert:

a 28% to a simple fraction

b $\frac{1}{6}$ to a percentage

c 12.3% to a decimal

4 A particular school has 40 female teachers and 24 male teachers. What percentage of teachers are female?

5 Evaluate each expression.

a $\frac{5}{6}+\frac{2}{3}$

b $3\frac{1}{5}+\frac{4}{15}$

C S F

HW HOMEWORK

9780170454520

PART C: PRACTICE

› Fractions and percentages revision

1 Jayden scored 19 out of 25 in a maths test. Convert this mark to a percentage.

____

2 Evaluate each expression.

a $\frac{4}{9} \times 2\frac{3}{4}$

____

____

b $\frac{14}{15} - \frac{4}{5}$

____

3 A cinema ticket price for a child will increase by 5% next year. What will be the new price if it is currently $14.60?

____

4 Find:

a $\frac{5}{8}$ of 1 day (in hours)

____

b 45% of 2.5 m (in cm)

____

5 The cost of a pair of shoes after a 7% discount is $74.40. Calculate:

a the original price of the shoes

____

____

b the discount.

____

C
S
F

PART D: NUMERACY AND LITERACY

1 Complete:

a Loss = ____________ – ____________

b Decreasing an amount by 13% is the same as finding ____________ % of the amount.

2 **a** Explain what an improper fraction is, giving an example.

____

____

b Complete: An improper fraction can be converted to a

____

3 A store buys a TV for $560 and sells it at a profit of $120. Find:

a the cost price ____________

b the selling price ____________

c correct to the nearest whole number, the profit as a percentage of cost price

____

____

4 How do you convert a decimal to a percentage?

____

____

HW HOMEWORK

6 FRACTIONS, DECIMALS AND PERCENTAGES

THIS SHEET WILL HELP YOU PRACTISE YOUR FRACTION, DECIMAL AND PERCENTAGE SKILLS.

1 Convert each percentage to a decimal.

a 17% ________ **b** 25% ________

c 9% ________ **d** 38% ________

e 87.5% ________ **f** 70% ________

g 4.1% ________ **h** 63.7% ________

i 0.5% ________ **j** 10.1% ________

2 Convert each decimal to a percentage.

a 0.15 ________ **b** 0.427 ________

c 0.08 ________ **d** 0.8 ________

e 0.015 ________ **f** 0.724 ________

g 0.69 ________ **h** 0.301 ________

i 0.007 ________ **j** 0.285 ________

3 Convert each percentage to a simple fraction.

a 76% ________ **b** 66% ________

c 5% ________ **d** 33% ________

e 40% ________ **f** 96% ________

g 62.5% ________ **h** 49% ________

i $16\frac{2}{3}\%$ ________ **j** 8.5% ________

4 Convert each fraction to a percentage.

a $\frac{3}{4}$ ________ **b** $\frac{13}{20}$ ________

c $\frac{3}{5}$ ________ **d** $\frac{7}{8}$ ________

e $\frac{19}{25}$ ________ **f** $\frac{7}{40}$ ________

g $\frac{2}{3}$ ________ **h** $\frac{38}{60}$ ________

i $\frac{11}{12}$ ________ **j** $\frac{11}{16}$ ________

5 Convert each decimal to a simple fraction.

a 0.6 ________ **b** 0.02 ________

c 0.18 ________ **d** 0.85 ________

e 0.07 ________ **f** 0.566 ________

g 0.72 ________ **h** 0.125 ________

i 0.048 ________ **j** 0.4 ________

6 Convert each fraction to a decimal.

a $\frac{5}{8}$ ________ **b** $\frac{9}{20}$ ________

c $\frac{13}{15}$ ________ **d** $\frac{3}{10}$ ________

e $\frac{1}{16}$ ________ **f** $\frac{4}{5}$ ________

g $\frac{34}{100}$ ________ **h** $\frac{67}{80}$ ________

i $\frac{6}{7}$ ________ **j** $\frac{7}{25}$ ________

7 Complete this table.

Fraction	Decimal	Percentage
		10%
	0.3	
$\frac{3}{4}$		
		20%
$\frac{1}{20}$		
	0.5	
		25%
		$66\frac{2}{3}\%$
$\frac{1}{8}$		
	0.6	

C S F

9780170454520

STARTUP ASSIGNMENT 7

THIS ASSIGNMENT REVISES STATISTICS, ESPECIALLY THE MEAN, MEDIAN AND MODE

PART A: BASIC SKILLS

/ 15 marks

1 Write $\frac{2}{3}$ as a percentage.

2 Find the perimeter of this semicircle to 2 decimal places.

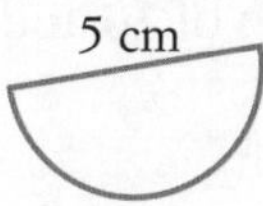

3 Complete: 1 m^3 = ______________ kL.

4 Bill and Ben share the weekly rent in the ratio 3 : 5. If Bill pays $105, what does Ben pay?

5 Expand $4(2x - y)$.

6 What is the angle sum of a quadrilateral?

7 What is the probability of a baby being born on a day beginning with T.

8 Find the area of this figure.

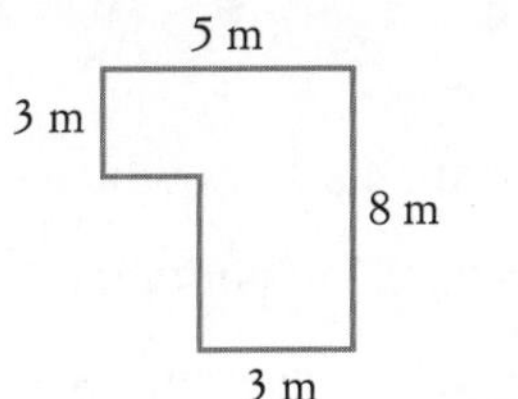

9 If $y = 4x + 5$, find y when $x = -2$.

10 Do the diagonals of a rectangle bisect each other?

11 Evaluate $\frac{2}{3} \div \frac{8}{15}$.

12 Find the length of the diagonal of this rectangle, to 2 decimal places.

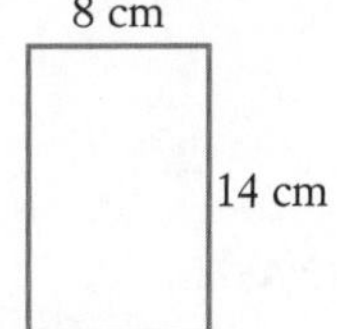

13 Use index notation to simplify $(2^5)^2$.

14 A scooter is bought for $180 and sold for $144. Calculate the loss as a percentage of the cost price.

15 Find the volume of this prism.

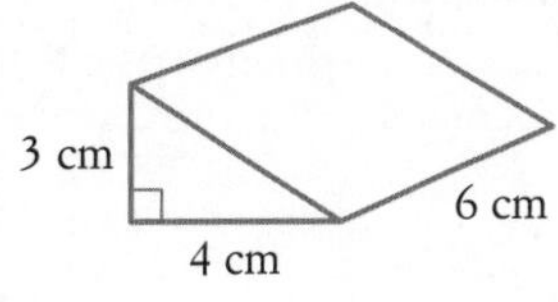

C S F

WS WORKSHEET

PART B: DATA

/ 25 marks

16 **a** Write these values in ascending order:

6 13 7 13 10 11 16 13 10

____

b Find the median. ______________________

c Find the mode. ______________________

d Find the range. ______________________

e Calculate the mean. ______________________

17 Claudia surveyed a group of students on the number of mobile phones owned at their homes. The data is illustrated on this dot plot.

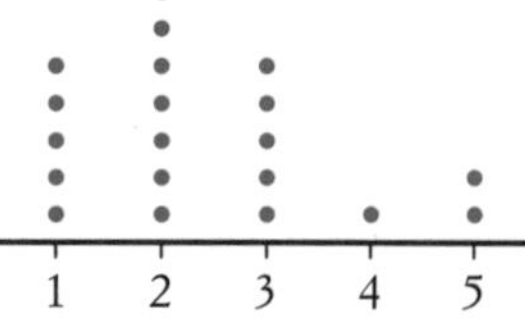

a How many students were surveyed?

____

b What was the mode?

____

c What was the highest data value?

____

d How many students had 3 mobile phones?

____

e What percentage of students owned more than 3 mobile phones?

____

18 Find the average of:

a 17 and 20

____

b 15, 24 and 30

____

19 The ages of children at a childcare centre are:

3 5 6 4 3 2 1 4 4 4 3 5

2 2 2 1 4 3 2 5 6 5 4 5

a How many children in the group? ______________________

b What fraction were 3 years old? ______________________

c What percentage (to one decimal place) were under 5? ______________________

d What was the mode? ______________________

e How many were this age? ______________________

f What was the lowest age? ______________________

20 Write 5 numbers that have a median of 10 and a mode of 12. ______________________

21 The speeds in km/h of a sample of cars are shown in this stem-and-leaf plot.

5	9
6	1 3 4 4 5 6 7 8 9
7	1 2 3 3 4 4 4 5 7 7
8	0 0 2 3

a How many cars were there? ______________________

b What was the mode? ______________________

c How many cars were moving faster than 80 km/h? ______________________

d What was the range? ______________________

e What fraction of cars were in the 60s stem?

____

PART C: CHALLENGE

Bonus / 3 marks

Jenna, Liz, Ally and Courtney are 4 sisters who work at the same bank. Liz earns $750 per week; Courtney earns $1200 per week. The average weekly wage of the 4 sisters is $3000. What could Jenna and Ally's wages be?

C S F

9780170454520

AS A GROUP OR INDIVIDUAL ACTIVITY, CLASSIFY THE DATA ACCORDING TO THE 3 GIVEN TYPES.

WS WORKSHEET

Categorical	Numerical: discrete	Numerical: continuous
Dress size	Volume of air in a balloon	Level of First Aid training
Intensity of a light	Marital status	Make of car
Goals scored by a hockey team	Method of travel to work	Film classification rating
Crowd size at a football match	Type of home lived in	Running time of an athlete
Height of a tree	Brand of toothpaste	Outcome of a coin toss
Exam mark as a whole percentage	Attitude towards a new national flag	Number of times a person streamed a film last week
Colour of eyes	Available memory on a computer	School population
Amount of chlorine in a pool	Traffic crossing a bridge in an hour	Loudness of a lawn mower
Month of birth	Word length of an essay	Year of birth
Floor area of an office	Mass of a truck	Sum of 2 dice rolled
Speed of a car	Price of an ice cream	Favourite radio station
Number of phones owned	Temperature of an oven	Amount of electricity used daily
Setting of an air conditioner	Distance lived from school	Exam grade (for example, B+)
A person's blood alcohol level	Pulse rate in beats/minute	House number
Day of week a person shops	Capacity of a car's fuel tank	Number of apps on phone
Number of bedrooms in home	Length of a phone call	Favourite sport

7 MEAN, MEDIAN, MODE

MATCH THE ANSWER TO EACH QUESTION TO ITS CORRECT VALUE, THEN USE THE MATCHED QUESTION NUMBERS AND LETTERS TO DECODE THE QUOTE ON THE NEXT PAGE.

PS PUZZLE SHEET

Find the **range.**

1 7, 5, 8, 2, 1, 9, 7

2 53, 74, 21, 66, 48

3 19, 74, 86, 50, 42, 63

4 3, 6, 3, 5, 4, 3, 6

5 61, 94, 75, 82, 89, 65

Find the **mode.**

6 5, 2, 7, 9, 2, 9, 3, 9

7 70, 75, 79, 71, 76, 75

8 2, 5, 7, 4, 2, 7, 9

9 32, 51, 76, 83, 40, 98

10 66, 48, 52, 57, 48, 51

Find the **median.**

11 1, 4, 10, 2, 6, 9, 7

12 49, 73, 65, 51, 34, 55, 87

13 85, 71, 90, 88, 60, 54, 69

14 3, 5, 3, 9, 7, 3, 4

15 6, 9, 2, 5, 9, 4, 8, 8

Find the **mean.**

16 5, 9, 3, 9, 7, 1, 4, 2

17 71, 84, 66, 52, 27

18 39, 42, 7, 56, 79, 34, 86

19 4, 7, 12, 9, 15, 13, 17

20 53, 89, 71, 83, 65, 29

A 11

B 55

C 8

D 9

E 65

F 53

G 2 and 7

H 7

I 4

L 3

M 60

N 71

O No mode

P 48

R 33

S 5

T 49

U 75

W 6

Y 67

9780170454520

‘

20	13	18	15	7	16	14	19	16	17

14	16

18	15	20

8	5	20	19	18	20	16	18

19	16	16	20	18

14	13

18	15	20

11	9	5	4	6

.

14	18

12	20	19	18	16

17	9	13	20	3

10	9	11	20	5

19	13	6

14	13	2	4	7	20	13	1	20

’.

Henry Chester, American author

PS PUZZLE SHEET

DATA CROSSWORD

PS PUZZLE SHEET

1 2 3 4 5 6 7 8 9 10 11 12 13 14 15 16 17 18 19 20 21 22 23 24 25 26 27 28 29 30 31 32

9780170454520

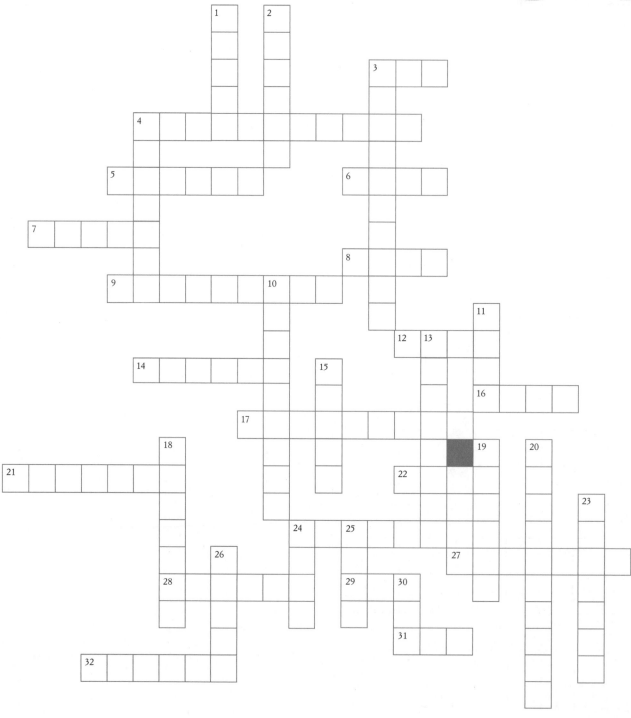

ACROSS

3 The mean of 2, 2, 6, 7, 9, 10

4 Data that is not numerical, for example, colour

5 To collect opinions or facts for statistical study

6 The median of 2, 2, 4, 6, 6, 10

7 The range of 2, 2, 6, 7, 9, 10

8 Something that influences a sample or survey unfairly

9 How often a value occurs in a data set

12 Stem-and-leaf or dot

14 A part of the population selected for survey

16 A measure of average calculated using all of the data values

17 Special frequency column graph

21 A frequency graph that looks like a mountain

22 A frequency polygon is a type of ______________ graph

24 The middle data value

27 An extreme data value

28 The range is a measure of ______________

29 The mode of 2, 2, 6, 7, 9, 10

31 This plot is a simple type of column graph

32 The mean is the sum of values divided by the ______________ of values

DOWN

1 Highest value – lowest value

2 Every item or person in the population has an equal chance of being selected in this type of sample.

3 The branch of mathematics that analyses data

4 Where many values in a data set are bunched

10 Data that is not categorical, for example, height

11 ______________ -and-leaf plot

13 Mean, median and mode are measures of ______________

15 The mode is the data value that occurs most ______________

18 To examine something by breaking it down into its parts

19 A survey of the whole population

20 All of the people or items being studied

23 Another word for mean

24 The most frequent value

25 Raw information

26 To find the median, you must first put the values in ______________

30 There is only one middle data value when there are an ______________ number of values

PS PUZZLE SHEET

7 STATISTICS 1

HI, MITCH HERE. THIS ASSIGNMENT REVISES IMPORTANT STATISTICAL SKILLS. IF YOU USE COMPUTERS, THEN YOU WILL SEE DATA.

Name:

Due date:

Parent's signature:

Part A	/ 8 marks
Part B	/ 8 marks
Part C	/ 8 marks
Part D	/ 8 marks
Total	/ 32 marks

HW HOMEWORK

PART A: MENTAL MATHS

Calculators not allowed

1 Complete: 8235 mL = ________________ L

2 A movie began at 2.20 p.m. and went for 2 hours 35 mins. What time did it finish?

____

3 Simplify $4ab - 2b + ba - 3b$. ________________

4 Round 183.495 to 2 decimal places.

5 Evaluate $\frac{5}{9}+\frac{1}{3}-\frac{1}{6}$ ________________

6 Find the value of x if the volume of this rectangular prism is 48 cm^3.

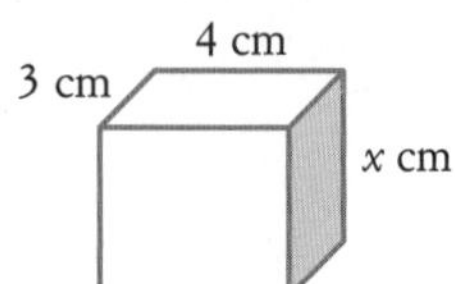

7 What is the probability of randomly selecting a male teacher from a school of 24 male and 40 female teachers?

____

8 One of the angles in an isosceles triangle is 130°. What are the sizes of the other 2 angles?

____

PART B: REVIEW

1 This dot plot shows the number of rainy days per week in Westvale over summer.

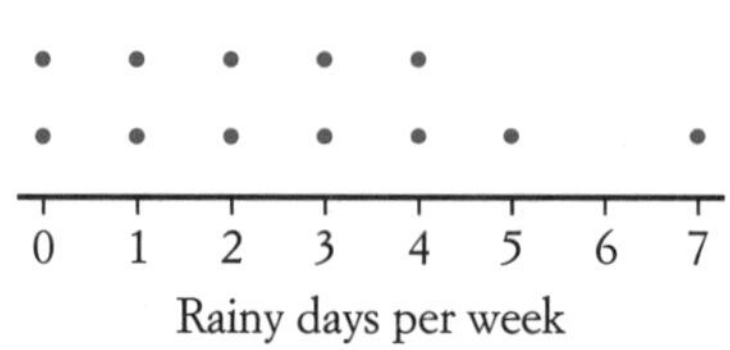

a What was the highest value? ________________

b How many weeks had 2 rainy days?

____

c How many weeks had no rainy days?

____

2 **a** Arrange these values in ascending order:

20 12 8 5 14 5 7 5 12

____

b Which value is an outlier? ________________

c Which value occured most often?

____

d What is the middle value?

____

C S F

9780170454520

3 This sector graph shows how Helena spends a typical day, in hours per activity. Calculate the size of the sector angle for 'Sleep'.

____

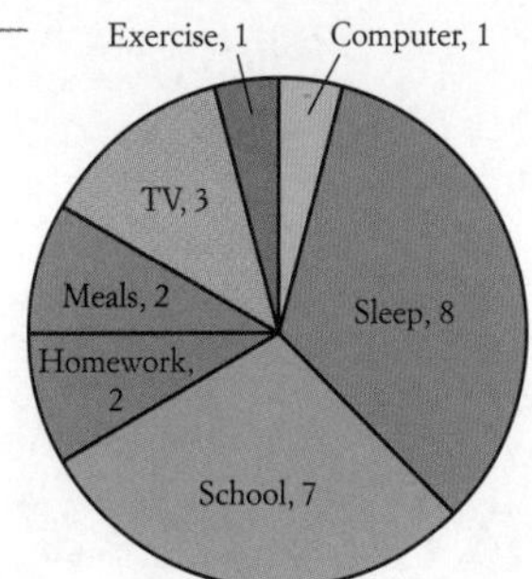

PART C: PRACTICE

› The mean, mode, median and range

1 The data below show the number of texts sent today by a group of students.

11 12 18 23 11 24 13 11 22 24 16

a How many students are in the group?

____

b What is the lowest value? _______________

c Find the mean correct to one decimal place.

____

d Find the mode. _______________

e Find the range. _______________

f Find the median.

____

2 These are people's scores out of 10 on a quiz:

7 4 8 5 3 8 5 8 3 6 7 4 9 9

a Find the mode.

____

b Find the median.

____

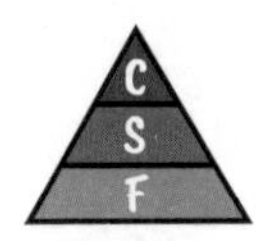

PART D: NUMERACY AND LITERACY

1 Which measure of centre:

a is calculated from all of the values in the data set?

____

b can have more than one value?

____

2 Complete: The range is the _______________ between the _________ value and the _________ value.

3 The set of data below is arranged in ascending order, with x representing an unknown number.

3 4 5 5 5 x 11 11 11 15

Find the value of x if:

a the median is 7

____

b the mean is 8

____

c the mode is 5

____

4 **a** If a set of data has an odd number of values, how many middle values are there?

____

b What is the median of an odd number of values?

____

____

____

HOMEWORK

7 STATISTICS 2

SO WHAT ARE THE 3 MEASURES OF CENTRE? THEY ALL BEGIN WITH THE LETTER M.

Name:

Due date:

Parent's signature:

Part A	/ 8 marks
Part B	/ 8 marks
Part C	/ 8 marks
Part D	/ 8 marks
Total	/ 32 marks

HW HOMEWORK

PART A: MENTAL MATHS

Calculators not allowed

1 Complete 4 : 5 = 24 : ______________.

2 Solve the equation $3m + 5 = 16$.

____

3 How many axes of symmetry has a parallelogram? ______________

4 Evaluate each expression.

a $5 \times 8 - 16 \div (-2)$ ______________

b $\frac{5}{9} \div \frac{20}{21}$

____

5 **a** Name this solid shape.

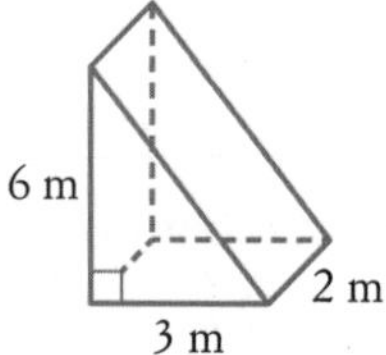

b Find its volume.

____

6 If a die is rolled, what is the probability that the number that comes up is a factor of 6?

____

PART B: REVIEW

1 For the data: 8 5 9 5 3 7 8 4 8 6 8 2

a find the mode

____

b find the median

____

c find the range

____

d show them on a dot plot

2 For the data:

46 11 4 15 4 11 7 4 12

a find the mode

____

b find the median

____

c find the mean correct to one decimal place

____

d find the outlier

____

C S F

9780170454520

PART C: PRACTICE

› Frequency tables
› Dot plots and stem-and-leaf plots

1 For the data shown in this frequency table:

Score x	Frequency f	fx
1	3	
2	4	
3	2	
4	6	
5	1	
Total		

a complete the table

b find the mean

c find the mode

2 For the data shown on this dot plot, find:

	6	7	8	9	10	11	12
Dots	2	5	3	2	5	1	1

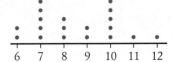

a the mode

b the median

3 For the data shown on this stem-and-leaf plot, find:

Stem	Leaf
0	7 8 9
1	0 3 3 5 5 7 7 8 9
2	3 4 4 4 6 6 8 8
3	0

a the range

b the median

c the mode

PART D: NUMERACY AND LITERACY

1 a If a set of data has an even number of values, how many middle scores are there?

b What is the median of an even number of values?

2 Complete: The __________ is calculated by __________ all of the data values, then __________ by the number of values.

3 12 students were surveyed on how many minutes they spent watching TV last night:

64 125 74 61 88 92

85 74 70 66 71 128

a (2 marks) Draw a stem-and-leaf plot for this data.

Stem	Leaf

b Find correct to 2 decimal places the mean.

c Find the range.

d Find the median.

HW HOMEWORK

9780170454520

7 STATISTICS 3

DOING STATISTICAL CALCULATIONS IS IMPORTANT, BUT IDENTIFYING BIAS AND UNFAIRNESS IN DATA IS EVEN MORE IMPORTANT.

Name:

Due date:

Parent's signature:

Part A	/ 8 marks
Part B	/ 8 marks
Part C	/ 8 marks
Part D	/ 8 marks
Total	/ 32 marks

PART A: MENTAL MATHS

Calculators not allowed

1 Convert $\frac{4}{9}$ to a decimal.

______

2 Expand $8(3a + 5)$. ______

3 What is the angle sum of a quadrilateral?

______

4 Evaluate each expression.

a $\sqrt{13^2 - 12^2}$

______

b $\frac{5}{8} - \frac{2}{5}$

______

5 Simplify $\frac{8be}{20bd}$. ______

6 Find the area of this shape.

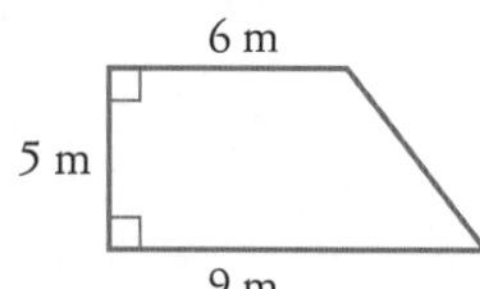

______

7 What is the probability of a person selected at random having a birthday in a month beginning with J? ______

PART B: REVIEW

1 This dot plot shows the number of rainy days per week in Westvale over summer. Find:

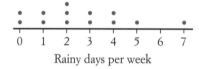

Dot plot (Rainy days per week): 0: 2 dots, 1: 2 dots, 2: 3 dots, 3: 2 dots, 4: 2 dots, 5: 1 dot, 6: 0 dots, 7: 1 dot

a the range ______

b the mode ______

c the median ______

d the mean correct to one decimal place.

______

2 For the data shown in this frequency table, find:

Score x	Frequency f
1	4
2	4
3	2
4	6
5	1

a the range ______

b the median ______

3 For the data shown on this stem-and-leaf plot, find:

Stem	Leaf
1	2 6 8 8 8 8 9
2	2 3 4 5 7 7 8
3	3 4 4 6

a the mean ______

______

b the median ______

______

9780170454520

PART C: PRACTICE

› Sampling and bias
› Comparing data sets

1 State whether each sample below is random or biased.

a Surveying the first 5 students to arrive at school about the number of children in their families ____________

b Selecting every 10th name drawn out of a box of all students' names for a school survey on healthy eating ____________

c Surveying a Year 8 class about their ideas for school sport ____________

2 Should a sample or a census be used to find:

a the most popular car colour in Australia?

____

b people's opinions on climate change?

____

3 For the data shown on this stem-and-leaf plot, find which city has:

Melbourne		Sydney
9 8 7	0	5 6 6 7 8 9
8 7 6 6 6 4 3 3 0	1	1 1 1 2 2 2 6 8 8 8 9
9 8 8 6 5 4 4 4 3	2	2 2 3 4 5
0	3	

a the wider range

____

b the higher mean

____

____

c the higher median

____

PART D: NUMERACY AND LITERACY

1 Write a word to describe:

a a high or low value that is much different to the others in a data set

____

b a survey of the whole population

____

2 Complete: A random sample is a sample in which every item or person in the population has an equal ____________ of being chosen.

3 Describe how you could select a random sample of 40 students for their views on the food sold at your school canteen.

____

____

____

4 Which measure of central tendency:

a is most affected by outliers?

____

b has the highest frequency in a data set?

____

5 A large group of customers at a busy shopping centre were interviewed on Saturday afternoon for a survey on Internet use. Explain why this sample may be biased.

____

HW HOMEWORK

8 STARTUP ASSIGNMENT 8

HI, I'M MS LEE. THIS ASSIGNMENT REVISES YOUR GEOMETRY SKILLS TO PREPARE YOU FOR THE CONGRUENT FIGURES TOPIC.

PART A: BASIC SKILLS

/ 15 marks

1 Place these values in ascending order:

$\frac{1}{8}$, 17%, 0.2, $\frac{7}{40}$

____

2 Find the volume of this prism.

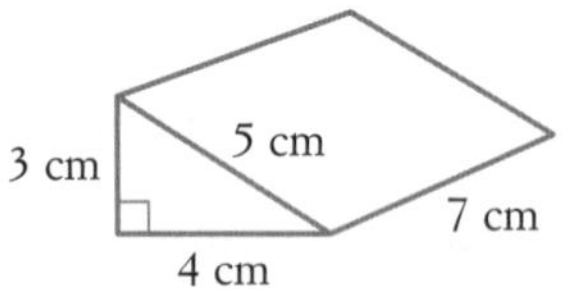

____

3 If (7, x, 25) is a Pythagorean triad, find x.

____

4 Evaluate $(-3)^3$. ______________________

5 Simplify $4d - 9 + 3d - 7$.

____

6 Draw a sector of this circle.

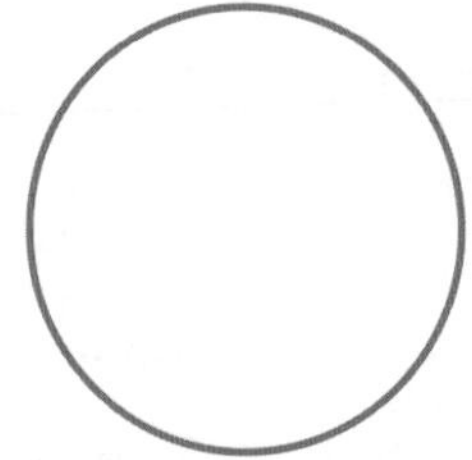

7 **a** What type of graph is this?

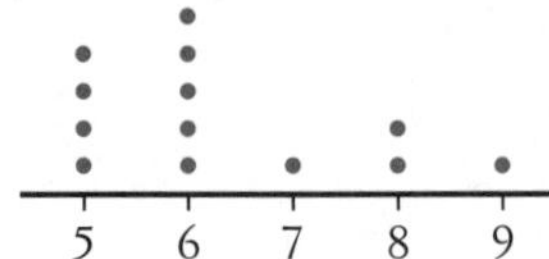

b What is the mode? ______________________

c What is the range? ______________________

8 Find 35% of 3 hours in hours and minutes.

____

9 Write the formula for the circumference of a circle with radius r. ______________________

10 Simplify $-4m \times 3mn$. ______________________

11 Convert $\frac{2}{9}$ to a decimal.

____

12 The ratio of dogs to cats at a pet show was 7 : 5. If there were 84 dogs, how many cats were there? ______________________

13 Find the volume of this cylinder correct to 2 decimal places

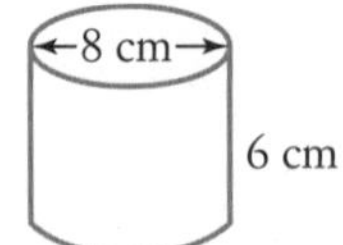

____

PART B: GEOMETRY

/ 25 marks

14 What is the most general quadrilateral with 4 equal angles? ______________________

15 Construct each triangle accurately on a sheet of paper.

a

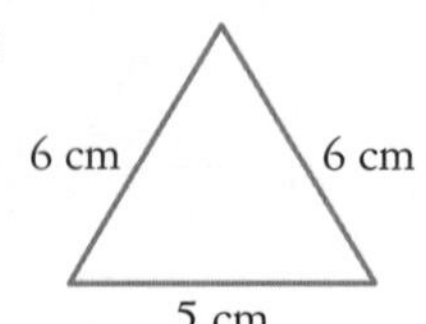

b

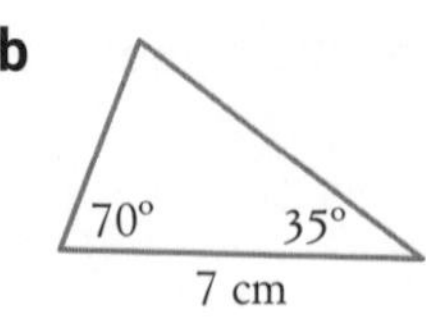

16 Translate this figure 4 units right and 4 units up.

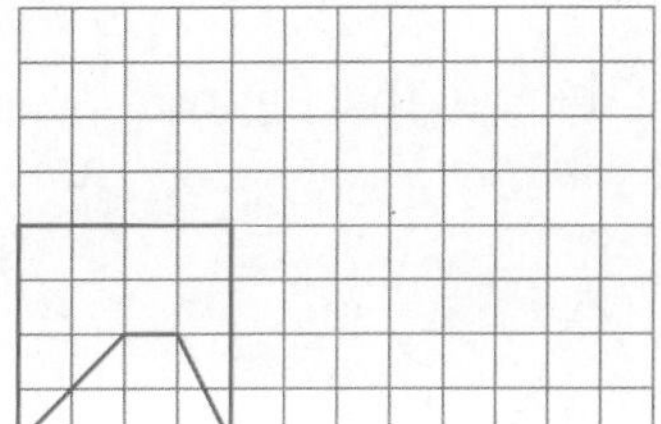

17 Are the opposite angles of a parallelogram equal? ______________________

18 Measure this angle.

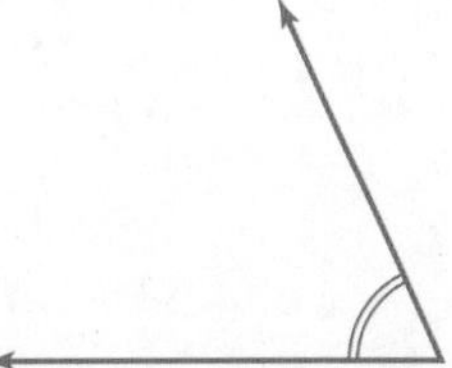

19 *ABCD* is a rhombus.

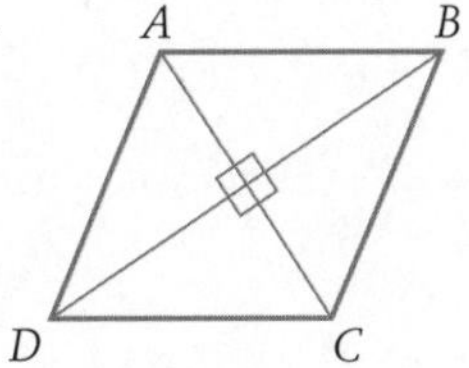

a What is the definition of a rhombus?

____

b How many axes of symmetry has a rhombus?

____

c Name any pair of parallel intervals.

____

d Name any pair of perpendicular intervals.

____

20 Rotate this figure 180° about point *P*.

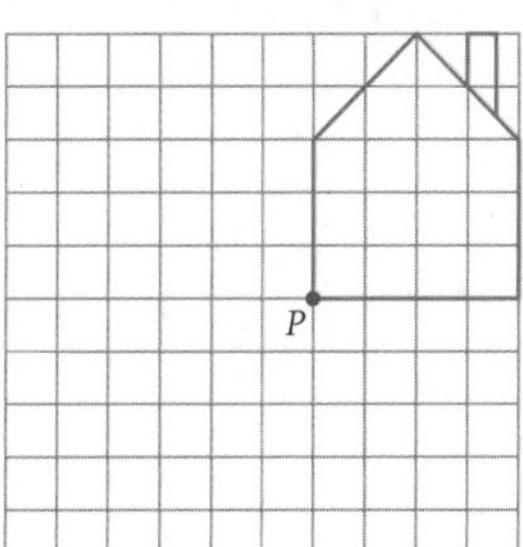

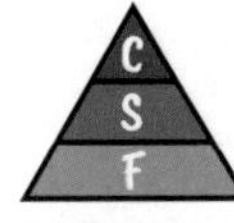

21 Find the value of each variable.

a

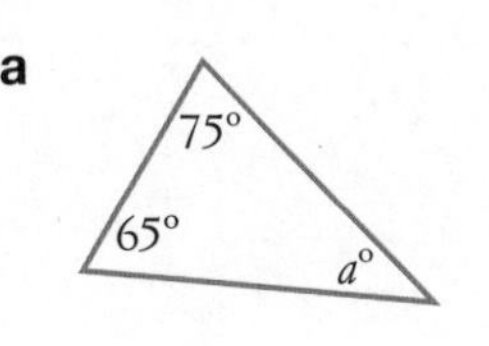

b

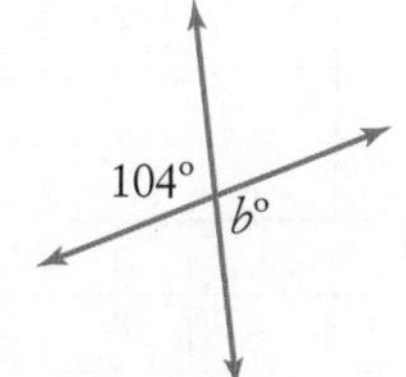

c

e

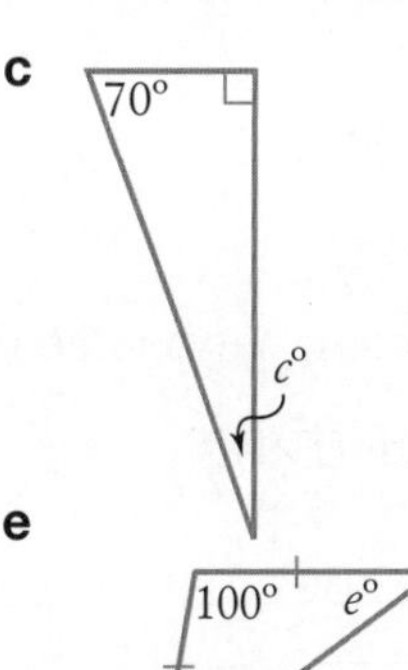

d

f

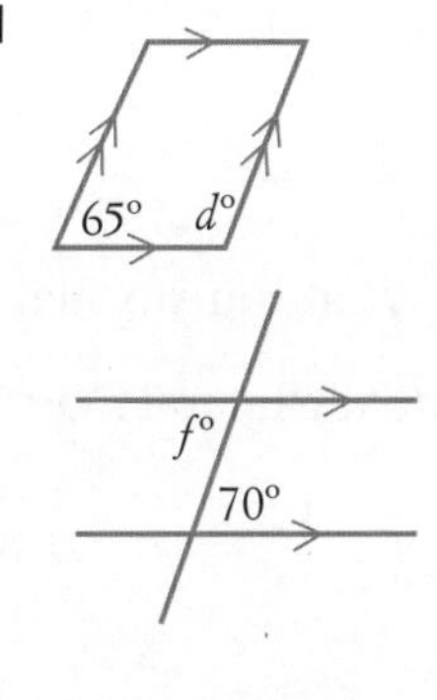

22 What type of angles are shown in question **21f**?

____

23 **a** Construct a right-angled triangle with 2 sides 5 cm long.

b Find, correct to one decimal place, the length of the triangle's hypotenuse:

i by measurement ______________________

ii by Pythagoras' theorem.

____

____

WS WORKSHEET

24 Reflect this figure across the given line.

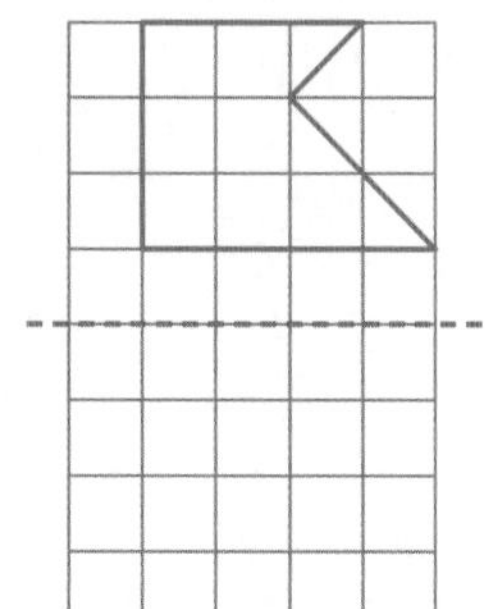

25 **a** What transformation has been performed on quadrilateral *PQRS* to create *UVWX*?

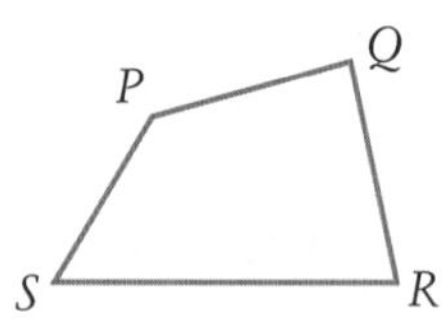

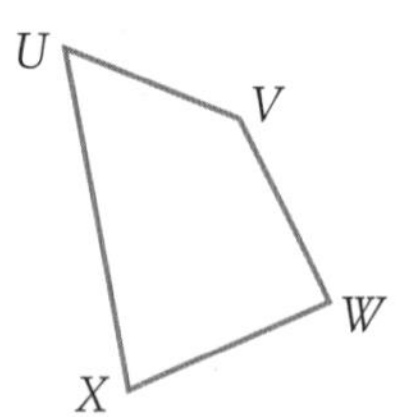

b Which side in *UVWX* matches with *PS*?

c Which angle in *UVWX* matches with $\angle Q$?

PART C: CHALLENGE

Bonus / 3 marks

How many triangles are there in this figure?

The answer is not 16.

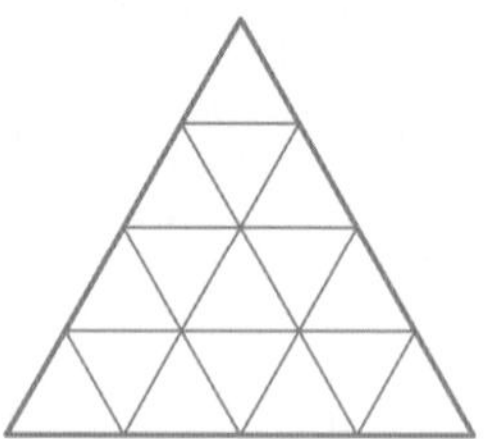

C
S
F

9780170454520

A PAGE OF CONGRUENT TRIANGLES 8

CONGRUENT MEANS 'IDENTICAL':
MATCHING EQUAL SIDES,
MATCHING EQUAL ANGLES.

WS WORKSHEET

By measurement and observation, find the congruent triangles. Colour each group a different colour. There are some triangles that are not part of a group.

8 TESTS FOR CONGRUENT TRIANGLES

FOR THE SAS TEST, THE 'INCLUDED' ANGLE MUST BE BETWEEN THE 2 SIDES.

Match the correct answer to each question to a letter shown, then use the matched question numbers and letters to decode the quote at the bottom of the next page.

Part A: Which test can be used to prove that each pair of triangles are congruent?

A AAS **V** RHS **D** SAS **W** SSS

1

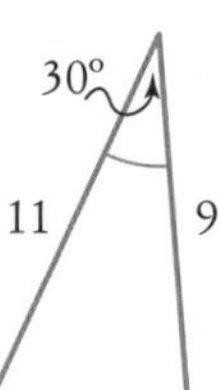

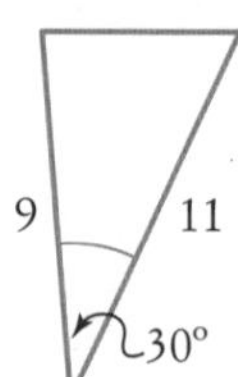

2

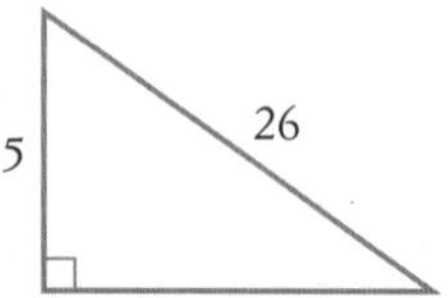

26
15

3

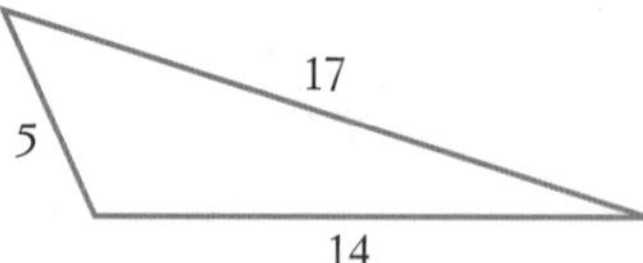

14
17
5

4

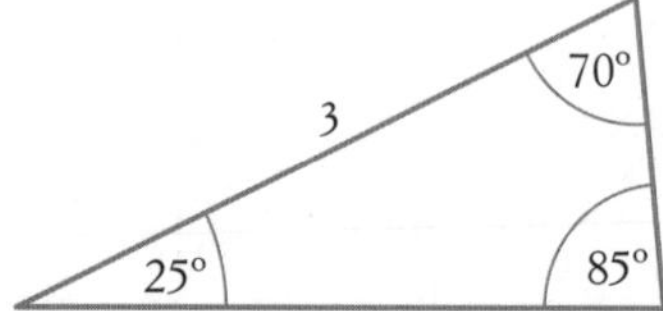

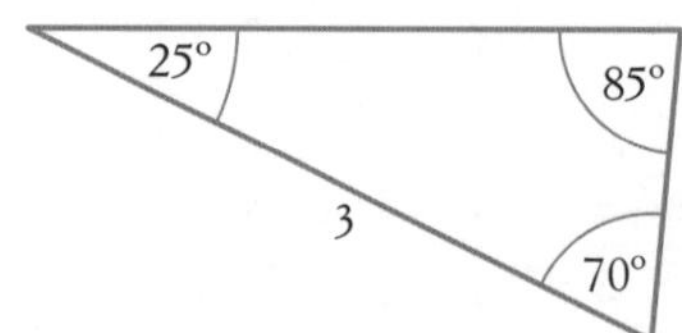

 9780170454520

Part B: Is each pair of triangles congruent or not congruent?

Question	5	6	7	8	9	10	11	12	13
Congruent	M	E	G	H	I	S	U	B	O
Not congruent	N	Y	R	P	F	T	L	C	K

5

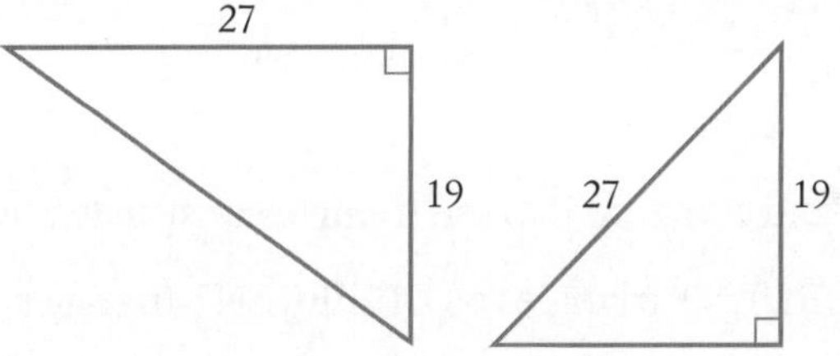

6

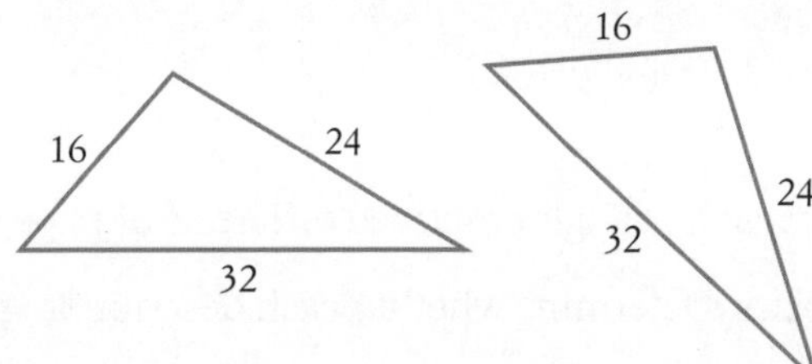

7

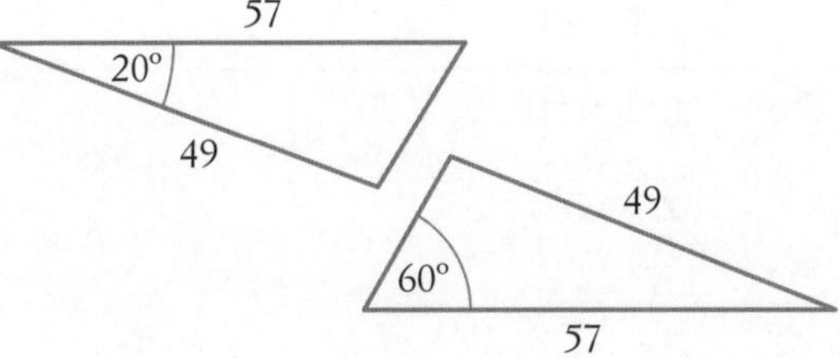

8

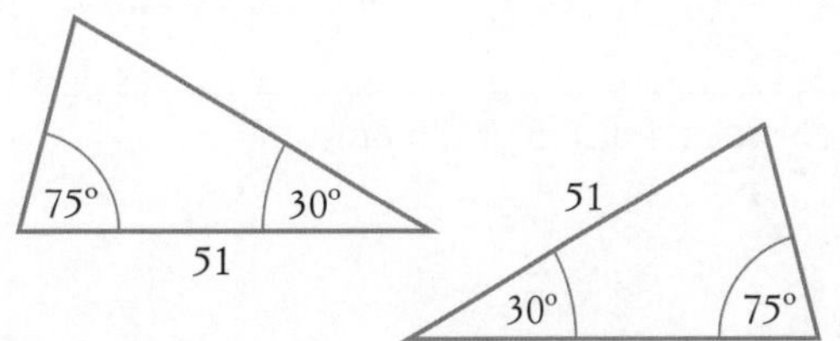

9

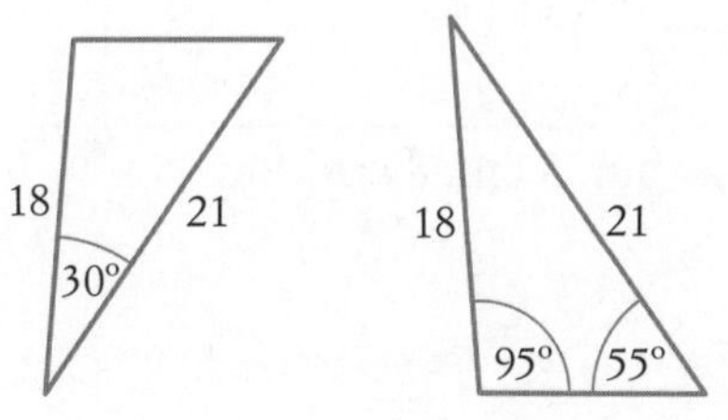

10

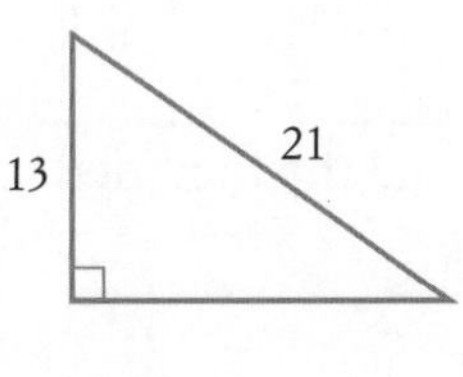

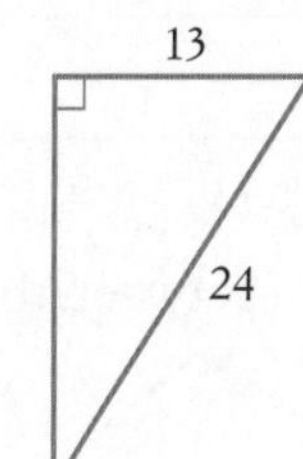

11

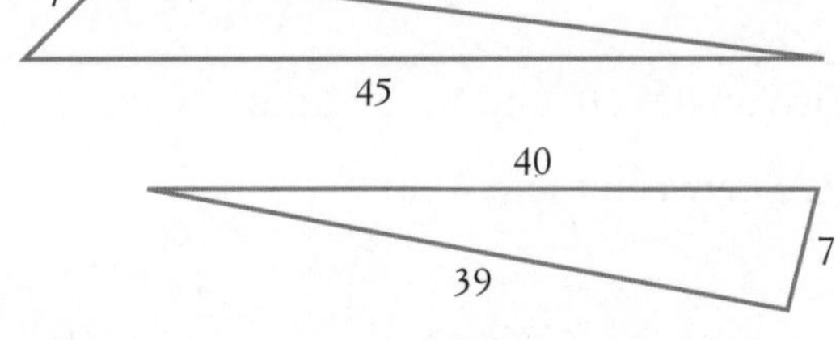

12

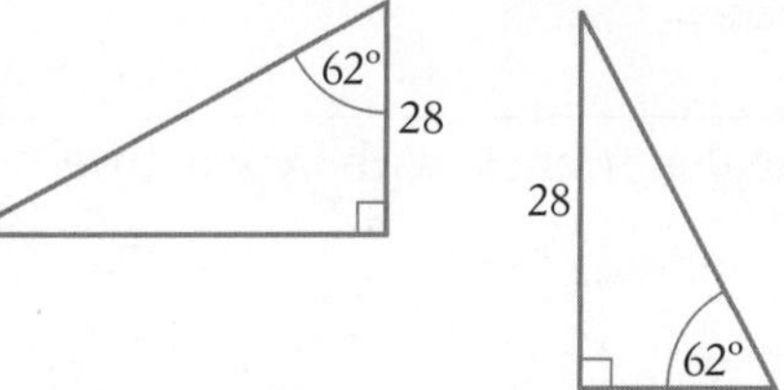

13

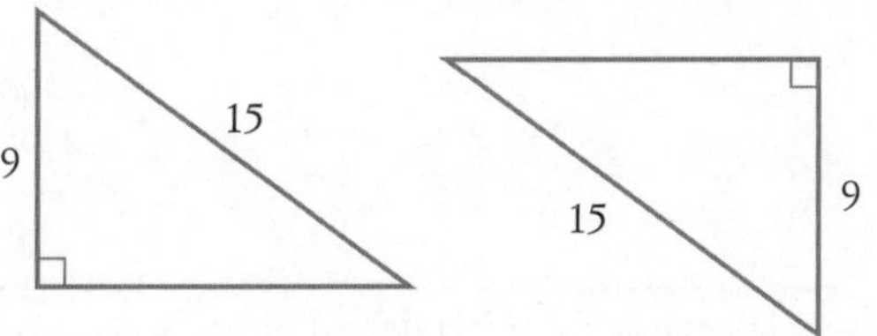

'The supreme happiness in life is ...

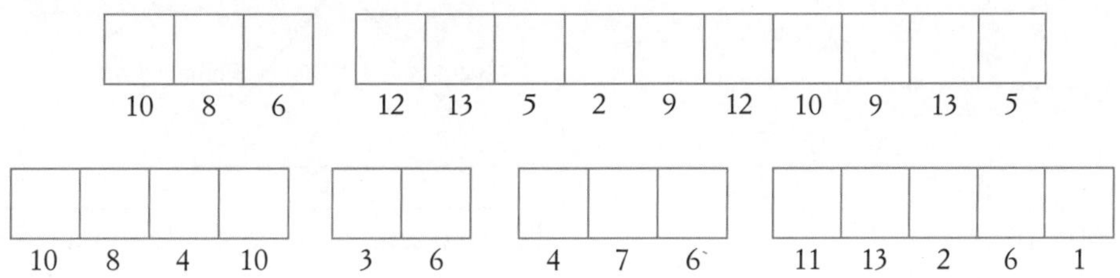

Victor Hugo, French novelist and poet

PS PUZZLE SHEET

WS WORKSHEET

8 CONGRUENT OR DIFFERENT TRIANGLES?

YOU NEED PAPER, SCISSORS AND GEOMETRICAL INSTRUMENTS FOR THIS ACTIVITY.

- Is it possible to draw 2 or more different triangles from the same description?
- How much information do you need about a triangle to be able to construct a congruent triangle?

Construct each triangle described on sheets of paper, cut it out and compare it to the triangle constructed by others in your class. Determine whether each description gives CONGRUENT triangles or DIFFERENT triangles.
Two of the triangles are IMPOSSIBLE: explain why.

1 Three sides 4 cm, 5 cm, 6 cm

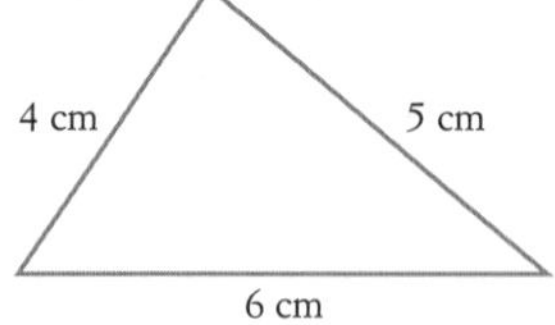

2 Three angles 50°, 60°, 70°

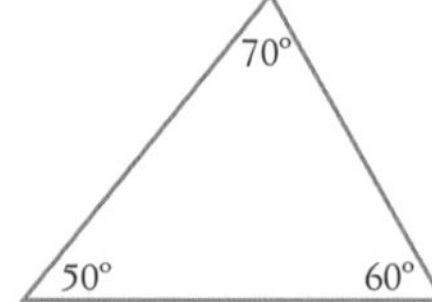

3 Two sides 4 cm and 6 cm and the included angle 30° (between the sides)

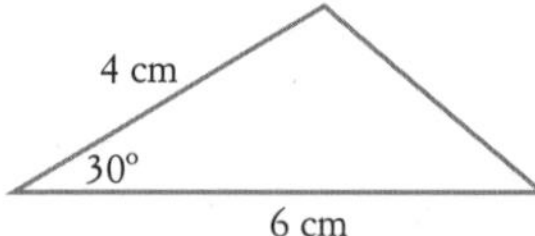

4 Three sides 4 cm, 3 cm, 8 cm

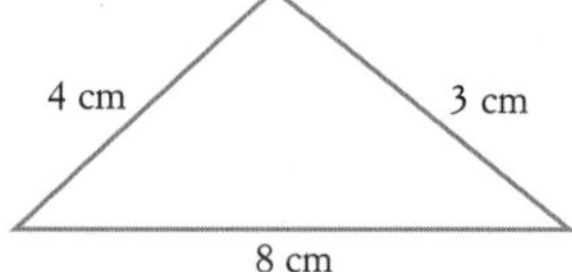

5 A right-angled triangle with hypotenuse 4 cm and one side 3 cm

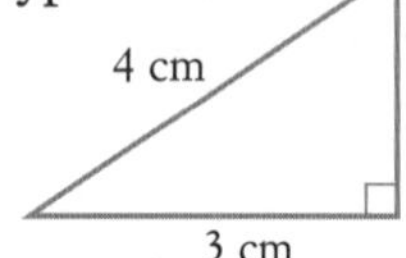

6 Two angles 60° and 75° and the side between them 4 cm

7 Two sides 5 cm and 3 cm and an angle 20° opposite the 3 cm side

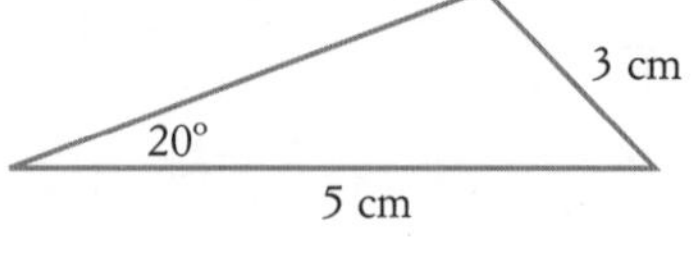

8 Isosceles triangle with equal sides 8 cm and equal angles 50°

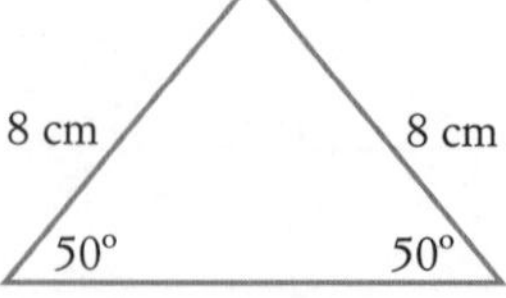

9 Three angles 60°, 70°, 60°

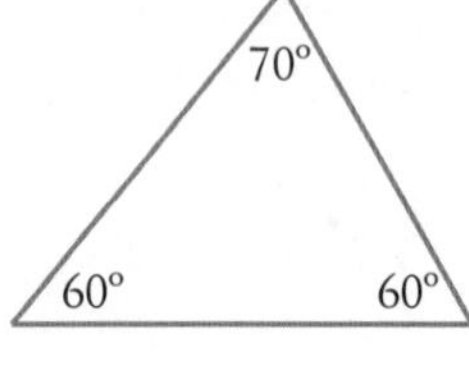

10 Two angles 30° and 50° and a side 6 cm opposite the 30° angle

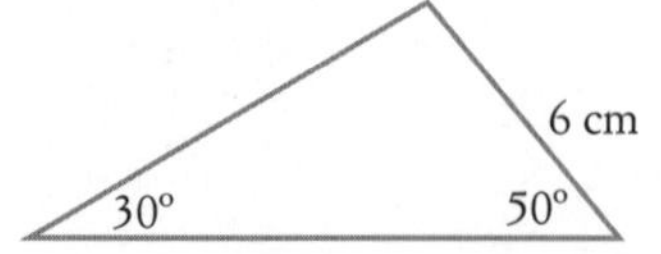

CONGRUENT FIGURES FIND-A-WORD 8


```
M C E G E T W B Y E H O L S H P W B R N N D K P K
D A Y L H L W M S X N B R R E U E Z C O V D S A Y
C E T M G N G U F P E E S R T H G I R I A D V R H
X D J C I N N N L Z I S P P D B T S D T L R R A D
C G I D H E A A A C X E O Q E O X W B A S E N L J
W O H F T I N I O T N J T P U Z J Q C L U D S L G
T C M O R O N N R D C C K B M Y L P N S B I X E R
O R P P G Q G G I T E E V P F I V T H N M S U L X
B Y A A A R V C M S O L R W Z K R E T A O U Y O P
H G I N U S U N I I P U V K O C P E R R H C M G T
Y D Q E S L S B P X P S V R P F F L P T R W U R I
E S N X A F X E P G O H I H F I H Q H U E V I A V
I T Q R X O O L S U S G N O I T A T O R S X Z M C
L H O Q H P T R A E I F U C N Y N T Q I O N E G O
C T P D I S D O M N T P F F T Y N J I P W N P L M
S L K M E Y O V A A E U W A A P E X F X Z T A F P
N S A T Z L I L X Y T I P S Q U A R E U Q M R P O
N G I Q O P R M T C X I V Z X J E B N H J V T A S
E M V M G Z I F U Q U A O D N Q D Y Q U K D I N I
Y K F H Y A V M S B U M F N H E E Q E P S N B G T
L A R E T A L I R D A U Q R E F L E C T I O N L E
J K N C J K T X N F U Q F B B R Q L B O I R E E J
C S L Q J X E C N M S S G T S P C T G O O K P I W
I N C L U D E D P Y G J K Q V W O S I N C N Y A D
T C U R T S N O C G I F Q R Q V J Z O W F A T I S
```

Find these words in the puzzle above. They are across, up and down, and diagonal, and can be backwards as well as forwards.

ANGLE
BISECT
COMPASSES
COMPOSITE
CONGRUENT
CONSTRUCT
DIAGONAL
HYPOTENUSE
IMAGE
INCLUDED
KITE
MATCHING
OPPOSITE
ORIGINAL
PARALLELOGRAM
PERPENDICULAR
QUADRILATERAL
RECTANGLE
REFLECTION
RHOMBUS
RIGHT
ROTATION
SIDE
SQUARE
SUPERIMPOSE
TEST
TRANSFORMATION
TRANSLATION
TRAPEZIUM
TRIANGLE
VERTEX

8 CONGRUENT FIGURES 1

SO WHAT ARE THE 4 TESTS FOR CONGRUENT TRIANGLES? EACH ONE CONTAINS 3 LETTERS.

Name:

Due date:

Parent's signature:

Part A	/ 8 marks
Part B	/ 8 marks
Part C	/ 8 marks
Part D	/ 8 marks
Total	/ 32 marks

PART A: MENTAL MATHS

Calculators not allowed

1 Complete:

a 45 L = ________ mL

b 1 m^3 = ________ cm^3

2 Expand $-4(2a - 3)$. ________

3 Round 84.208 to the nearest hundredth.

____

4 Evaluate 75% of $80. ________

5 Find the area of this shape.

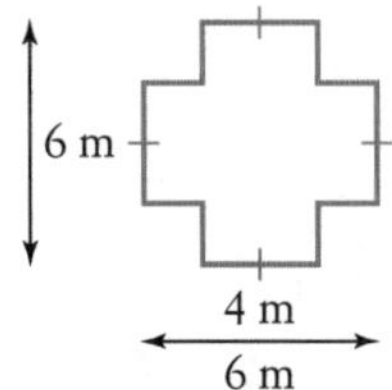

6 How many hours and minutes are there between 4.50 p.m. and 9.35 p.m.?

____

7 The ratio of boys to girls at the pool was 3 : 4. If there were 24 boys, how many girls were there?

____

PART B: REVIEW

1 (2 marks) Classify this triangle by sides and by angles.

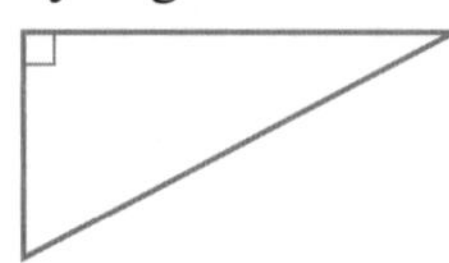

2 Name each type of transformation shown.

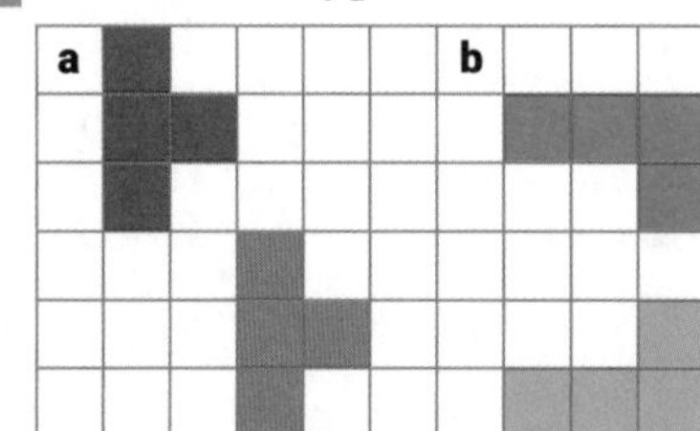

3 **a** Draw a rhombus.

b How many axes of symmetry has a rhombus?

____

c Write one property of the diagonals of a rhombus.

____

4 Reflect the figure across line AB, then rotate it 90° clockwise about C.

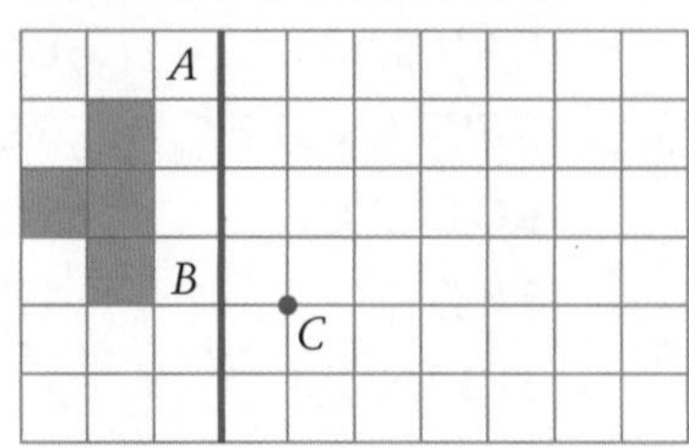

C S F

HW HOMEWORK

9780170454520

PART C: PRACTICE

› Congruent figures
› Tests for congruent triangles

1 These 2 trapeziums are congruent.

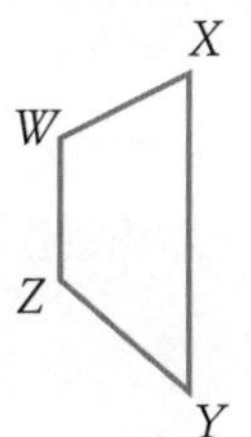

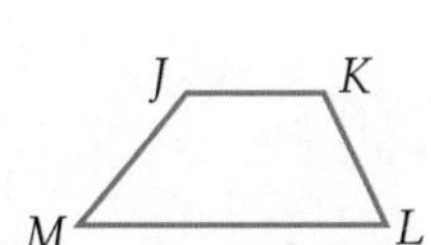

a Which transformation has been performed on *WXYZ* to create *JKLM*?

____

b Which side in *JKLM* matches with *ZY*?

____

c Which angle matches with $\angle X$? ________

d Complete using the correct order of vertices:

$WXYZ \equiv$ ____________________

2 (2 marks) Name 2 pairs of congruent triangles.

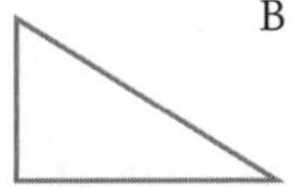

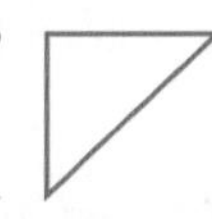

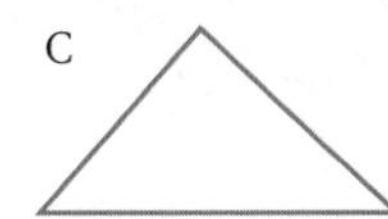

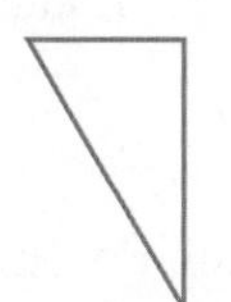

F

____

3 **a** Which 2 triangles are congruent?

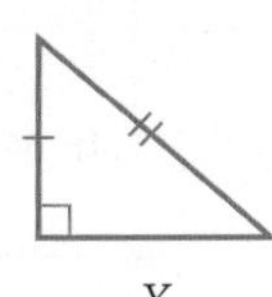

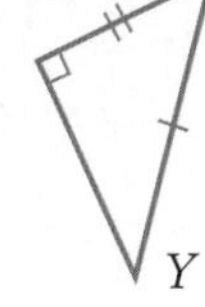

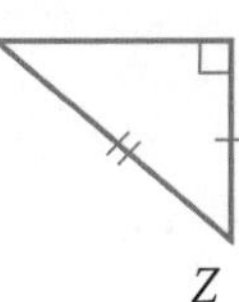

____

b Which congruence test proves this?

____

PART D: NUMERACY AND LITERACY

1 What does it mean to say that 2 shapes are congruent?

____

____

2 For the congruent triangle tests SAS and RHS:

a what type of angle is the 'A'?

____

b what does the 'R' stand for?

____

c what does the 'H' stand for?

____

3 What does the symbol '≡' stand for exactly?

____

4 Which congruence test proves that these 2 triangles are equal?

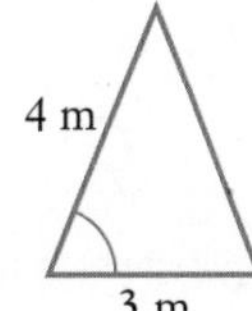

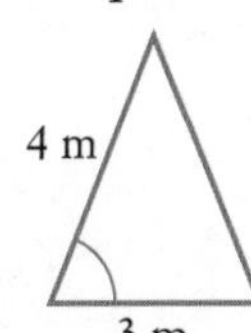

____

5 Draw an example of 2 congruent triangles:

a using the SSS test

b using the AAS test.

HW HOMEWORK

C S F

8 CONGRUENT FIGURES 2

A CLUE FOR PART A: A HECTARE IS THE AREA OF A SQUARE THAT'S 100 M BY 100 M.

Name:

Due date:

Parent's signature:

Part A	/ 8 marks
Part B	/ 8 marks
Part C	/ 8 marks
Part D	/ 8 marks
Total	/ 32 marks

PART A: MENTAL MATHS

Calculators not allowed

1 Express 40 minutes as a fraction of 2 hours.

---

2 Complete: 1 ha = ________ m^2

3 How many hours and minutes between 14:40 and 19:30?

---

4 Find y.

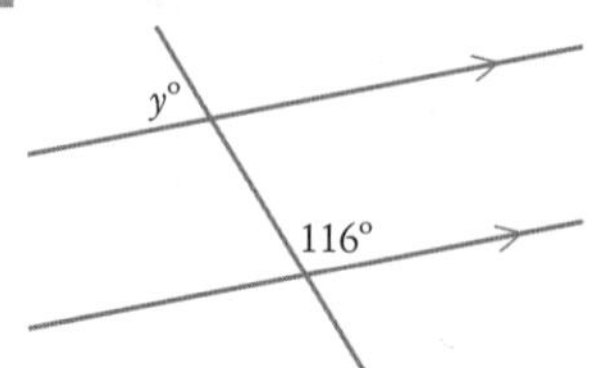

5 Evaluate each expression.

a 1.2×9

---

b $\frac{3}{8} \div \frac{5}{16}$

---

6 Factorise $18a + 24$. ________

7 Find the volume of this solid.

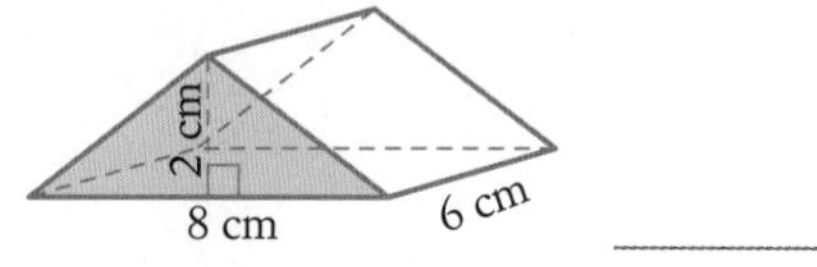

PART B: REVIEW

1 Name each quadrilateral.

a

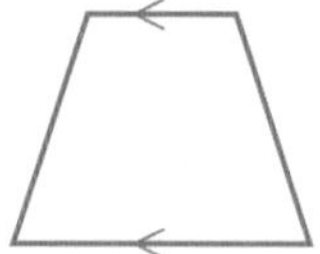

b

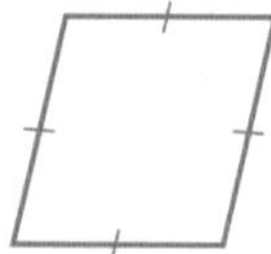

---

2 These 2 triangles are congruent.

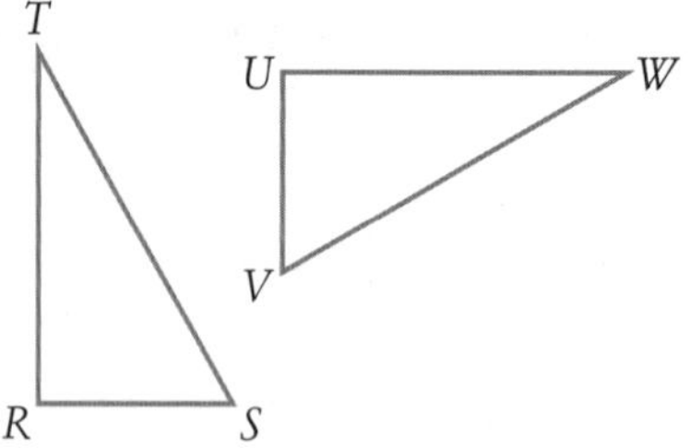

a Which side in $\triangle TSR$ matches with UW?

---

b Which angle matches with $\angle V$? ________

c Complete: $\triangle TSR \equiv \triangle$ ________

3 **a** What is a rectangle? ________

---

b What order of rotational symmetry has a rectangle? ________

c Write one property of the diagonals of a rectangle. ________

C S F

9780170454520

PART C: PRACTICE

› Constructing triangles
› Properties of quadrilaterals
› Constructing parallel and perpendicular lines

1 (3 marks) Use geometrical instruments to construct each diagram.

a A triangle with sides of length 3 cm, 4 cm and 5 cm.

b A line through X perpendicular to the line shown.

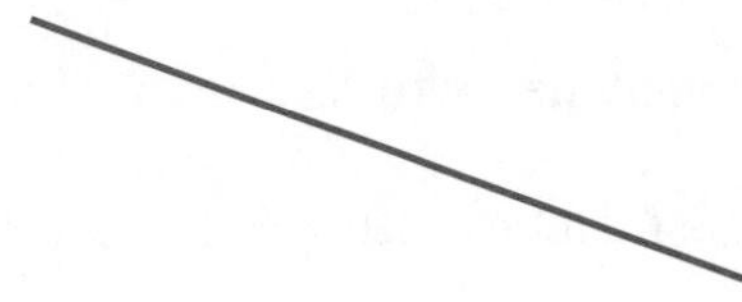

2 The diagonals of a rhombus divide the rhombus into 4 congruent triangles.

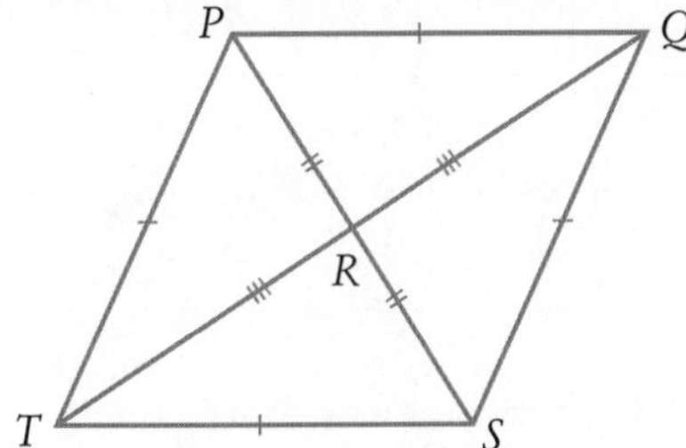

a Which test proves that the 4 triangles are congruent? ____________

b Hence find the sizes of the 4 angles at point R. ____________

c Mark $\angle TPR$ and the 3 angles that are equal to it.

d Are the diagonals of a rhombus equal?

____

e Complete: The diagonals of a rhombus ________ the angles of the rhombus

PART D: NUMERACY AND LITERACY

1 What are perpendicular lines?

____

____

2 Name the 2 tests for congruent triangles that involve 2 pairs of matching sides and one pair of matching angles.

____

3 (2 marks) Circle all quadrilaterals that have diagonals that cross at right angles: square, rectangle, trapezium, kite, rhombus, parallelogram.

4 What is a parallelogram?

____

____

5 Which test proves that these 2 triangles are congruent?

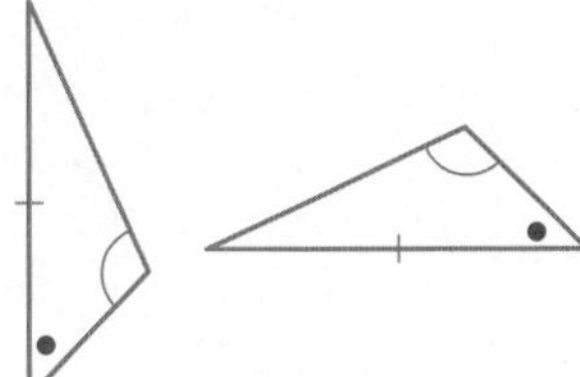

____

6 **a** Which quadrilaterals have 4 equal angles?

____

____

b What is the size of each angle? ____________

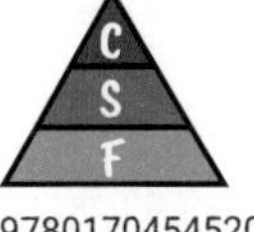

8 CONGRUENT FIGURES REVISION

Name:

Due date:

Parent's signature:

Part A	/ 8 marks
Part B	/ 8 marks
Part C	/ 8 marks
Part D	/ 8 marks
Total	/ 32 marks

A CLUE FOR PART B: A PARALLELOGRAM HAS OPPOSITE SIDES PARALLEL.

PART A: MENTAL MATHS

Calculators not allowed

1 Complete:

a 6 ML = ________ L

b 3.8 m^3 = ________ L

2 What is the time 3 hours and 10 minutes before 2.55 p.m.?

____

3 Simplify $6ab - 2ab + ba$. ________

4 Evaluate each expression.

a $\frac{5}{8} + \frac{1}{4}$

____

b $4.5 + 0.57 + 11.84$

____

5 Find the area of this shape.

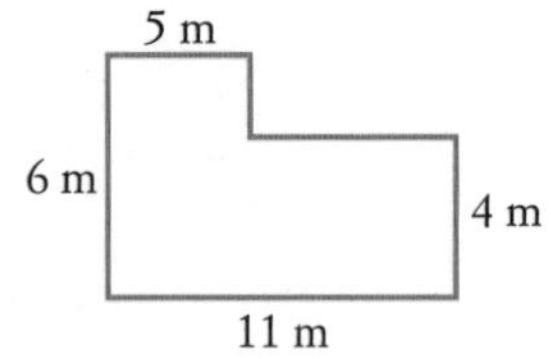

6 Decrease $60 by 30%.

____

PART B: REVIEW

1 **a** Which test proves that these 2 triangles are congruent? ________

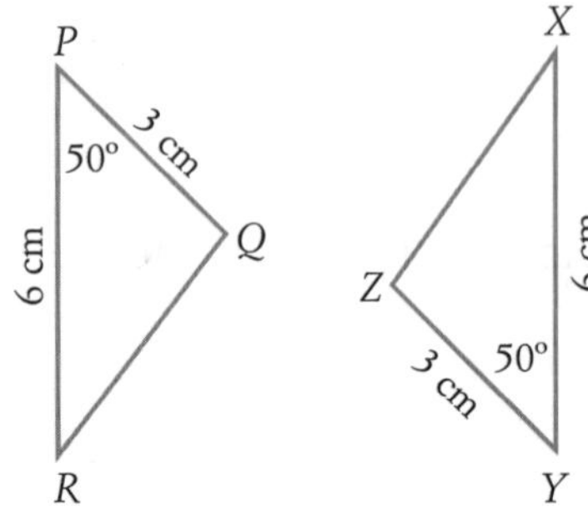

b Which side matches with XZ? ________

c Which angle matches with $\angle R$? ________

d Complete: $\triangle PQR \equiv \triangle$ ________

2 **a** What is a square?

____

____

____

b What order of rotational symmetry has a square? ________

c Is a square a special type of parallelogram?

____

d Complete: The diagonals of a square ________ each other at ________.

C S F

9780170454520

HW HOMEWORK

PART C: PRACTICE

› Congruent figures revision

1 (3 marks) Draw:

a a triangle with angles of 55° and 25° joined by a side of length 4 cm.

b a line through X parallel to the line shown.

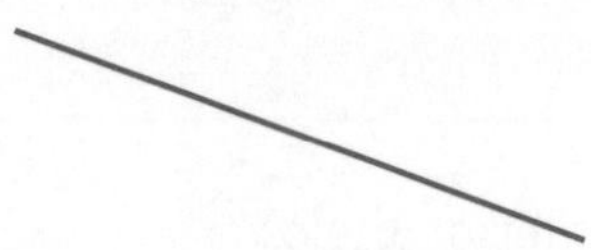

2 The axis of symmetry of an isosceles triangle divides it into 2 congruent triangles.

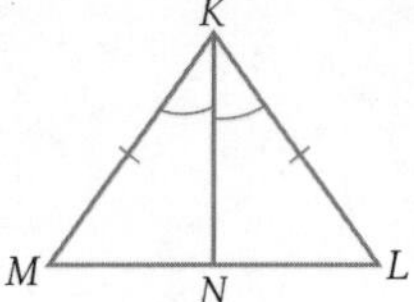

a Which test proves that the 2 triangles are congruent? ________________

b Why is $\angle KNM = \angle KNL$?

____

____

c What is the size of $\angle KNM$ and $\angle KNL$?

____

d Which angle matches with $\angle M$? ________

e Complete: An isosceles triangle has 2 equal ________, which are ________ the 2 equal sides.

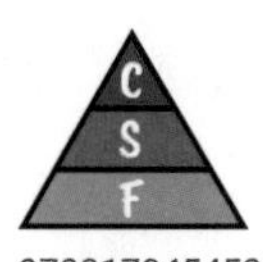

PART D: NUMERACY AND LITERACY

1 (2 marks) Circle all quadrilaterals that have 2 pairs of equal opposite sides: square, rectangle, trapezium, kite, rhombus, parallelogram.

2 Which quadrilaterals have 4 equal sides?

____

3 The axis of symmetry of a kite divides it into 2 congruent triangles.

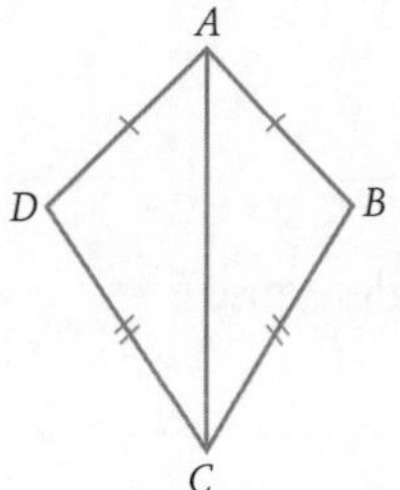

a Which test proves that the 2 triangles are congruent? ________________

b Which angle matches with $\angle CAB$? ________

c If $\angle DAB = 85°$ and $\angle D = 105°$, find the size of $\angle DCB$.

____

____

4 Draw an example of 2 congruent triangles:

a using the RHS test

b using the SAS test

HW HOMEWORK

9 STARTUP ASSIGNMENT 9

WS WORKSHEET

LET'S GET READY FOR PROBABILITY, THE MATHEMATICS OF CHANCE.

PART A: BASIC SKILLS

/ 15 marks

1 Evaluate:

a 8^3 ______

b $(-2)^5$ ______

2 Which is larger: $\frac{3}{5}$ or 66%?

______

3 Are the diagonals of a rectangle equal?

______

4 Calculate this circle's circumference to 2 decimal places. ______

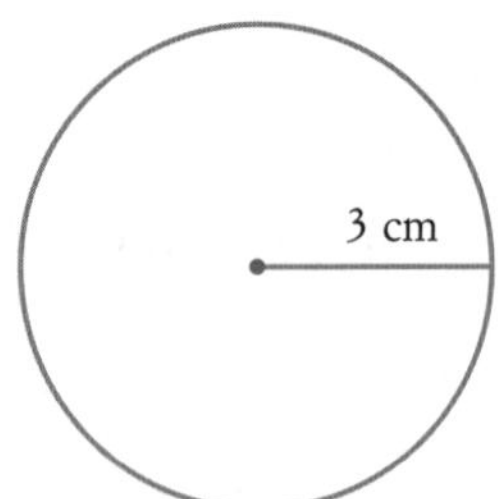

5 Write an algebraic expression for the number 4 less than y. ______

6 Simplify $\frac{2}{5} - \frac{1}{4}$. ______

7 Find r.

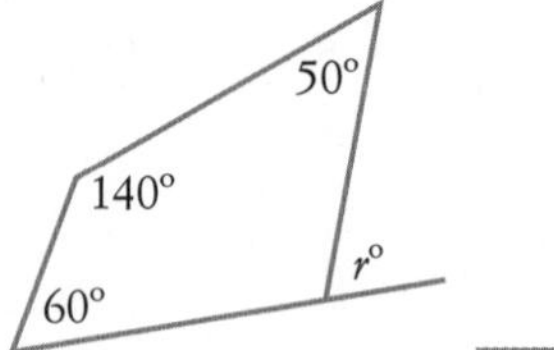

______

8 What are complementary angles?

______

______

9 Simplify $\frac{40pr}{-2pr^2}$. ______

10 Is a person's height an example of categorical data or numerical data? ______

11 What fraction of 3 hours is 40 minutes?

______

12 Mark a pair of co-interior angles.

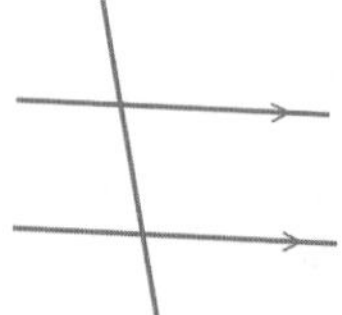

13 Factorise $32xy^2 - 20xy$.

______

14 Write Pythagoras' theorem for this triangle.

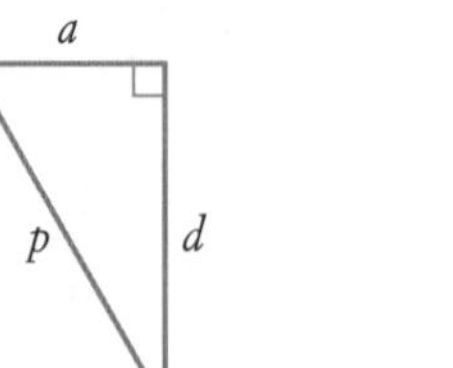

______

PART B: PROBABILITY

/ 25 marks

15 Simplify:

a $\frac{16}{40}$ ______

b $\frac{30}{48}$ ______

c $\frac{9}{16} + \frac{3}{16}$ ______

C S F

9780170454520

16 How many:

a consonants are there in the alphabet? ______

b outcomes are possible for the result of a soccer match? ______

c prime numbers are less than 10? ______

17 Convert to a percentage:

a $\frac{18}{20}$ ______

b 0.12 ______

18 What type of event has a probability of 100%? ______

19 Simplify:

a $1-\frac{3}{8}$ ______

b $1-\frac{7}{9}$ ______

c $1 - 0.982$ ______

20 What is the probability not rolling a 1 or 5 on a die? ______

21 What is the probability of selecting a month of the year that begins with the letter J? ______

22 At the cinema, 44% of the people were adults. What percentage were children? ______

23 Convert to a simple fraction:

a 36% ______

b 0.65 ______

24 Anton selects a lolly at random from a bag of 40 lollies that contains 10 green, 8 red and 12 white lollies. The rest are yellow.

a What is the probability that the selected lolly is yellow? ______

b Which colour has the highest chance of being selected? ______

25 (5 marks) Arrange these events in order, from the most likely to the least likely:

A Getting a tail when flipping a coin.

B A tennis player beating another higher ranked player.

C You have something to drink later today.

D It snows at your house this month.

E You will be home by 4 p.m. today.

______

26 What are the possible outcomes for the colour displayed on a traffic light?

______

______

PART C: CHALLENGE

Bonus / 3 marks

A $4 \times 4 \times 4$ cube is made up of 64 cubes stuck together. If this cube is painted red on all faces, then how many of the small cubes will have:

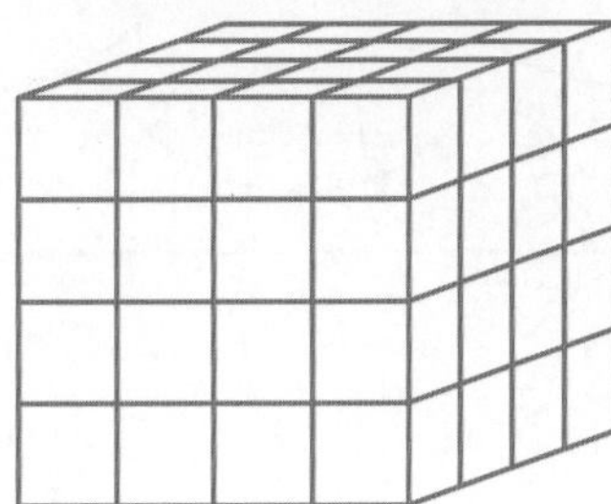

a one red face? ______

b 2 red faces? ______

c 3 red faces? ______

d no red faces? ______

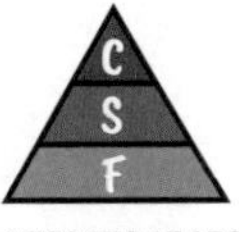

9 VENN DIAGRAMS

VENN DIAGRAMS ARE USED TO SORT ITEMS INTO CATEGORIES.

WS WORKSHEET

Enlarge this page for best results. Use each Venn diagram to classify the students (or half the students) in your class (write their initials), then answer the questions.

A

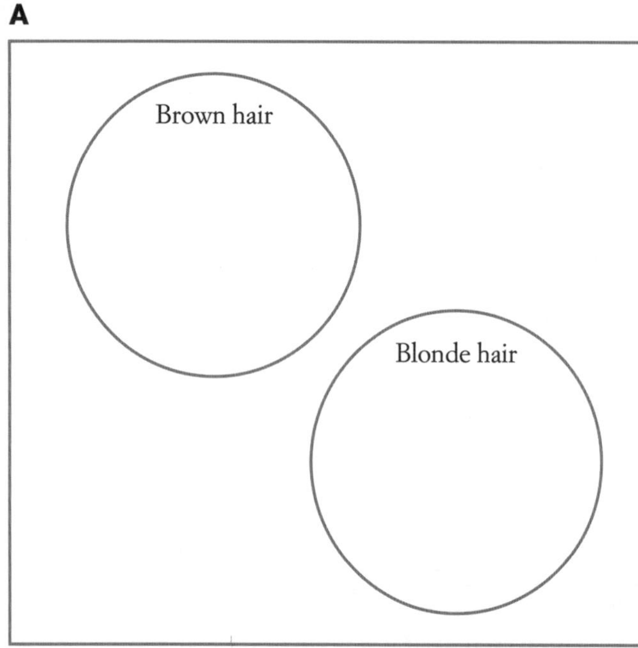

B

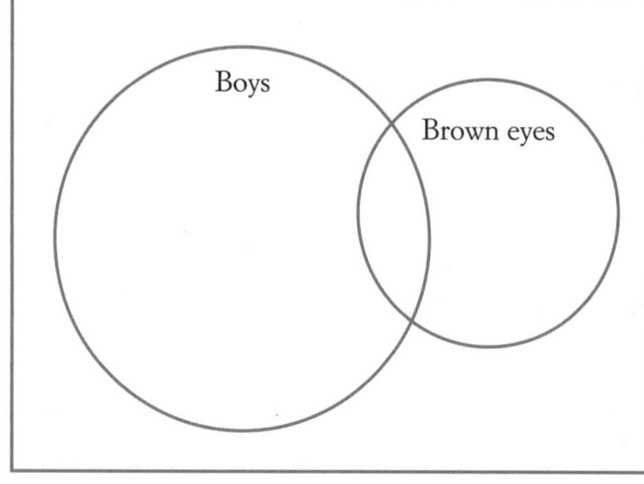

C

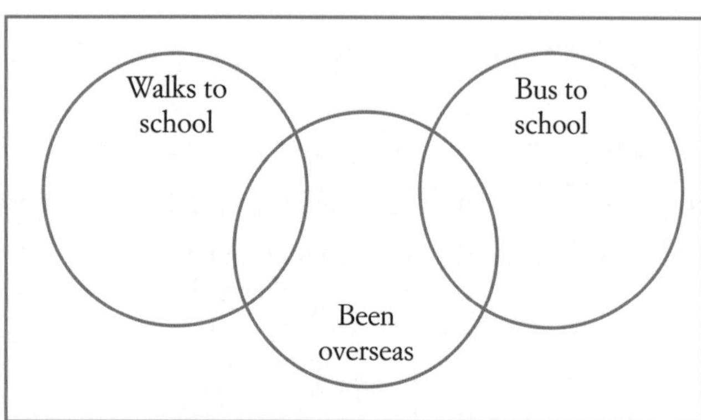

D

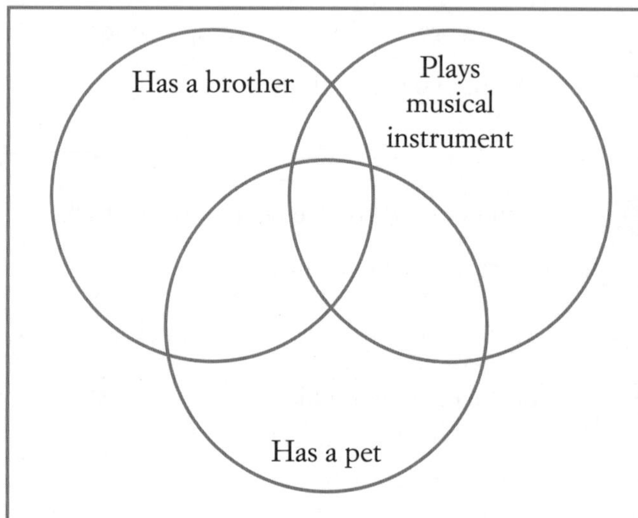

9780170454520

E

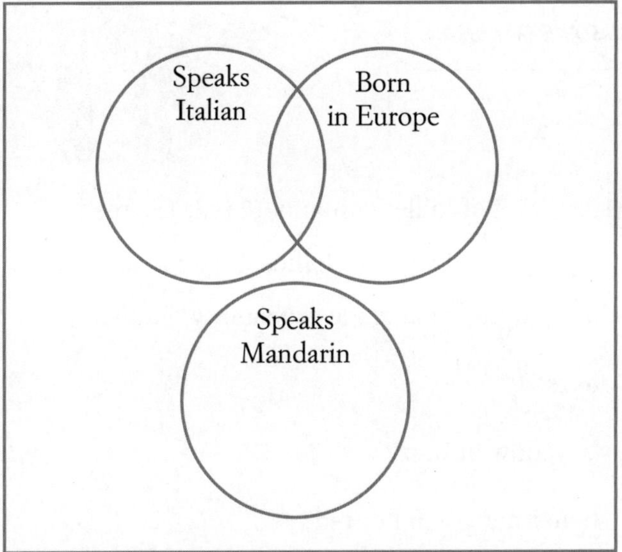

F

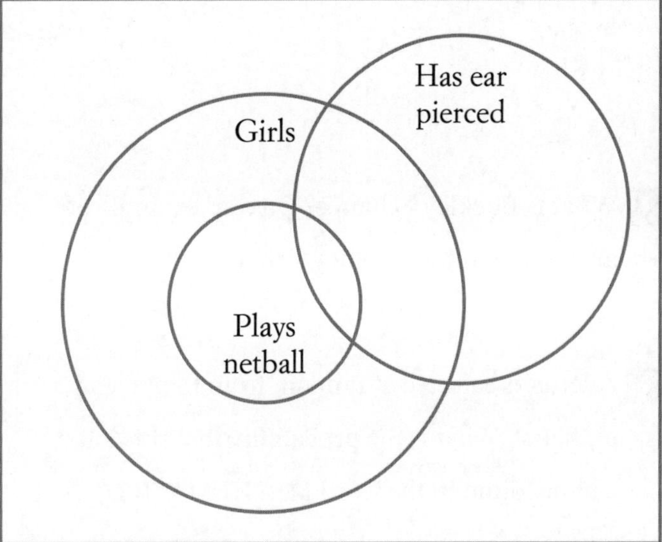

How many students:

a have blonde or brown hair? ____________________

b are boys with brown eyes? ____________________

c are girls? ____________________

d haven't travelled overseas? ____________________

e walk or ride a bus to school? ____________________

f ride a bus to school and have travelled overseas? ____________________

g have a brother and play an instrument? ____________________

h have a brother, pet and play an instrument? ____________________

i don't have a brother but have a pet? ____________________

j were born in Europe but don't speak Italian? ____________________

k speak Italian but weren't born in Europe? ____________________

l play netball? ____________________

m are boys with pierced ears? ____________________

n play netball and don't have pierced ears? ____________________

9 PROBABILITY REVIEW

BUCKLEY'S CHANCE IS AN OLD AUSTRALIAN SLANG EXPRESSION.

WORKSHEET

1 What is Buckley's chance? Give an example of an event that has Buckley's chance.

2 A letter is selected at random from the alphabet. What is the probability that this letter can be found in the word MATHEMATICS?

3 What is the probability that this spinner lands on:

a cerise?

b beige or violet?

c a colour that is not orange?

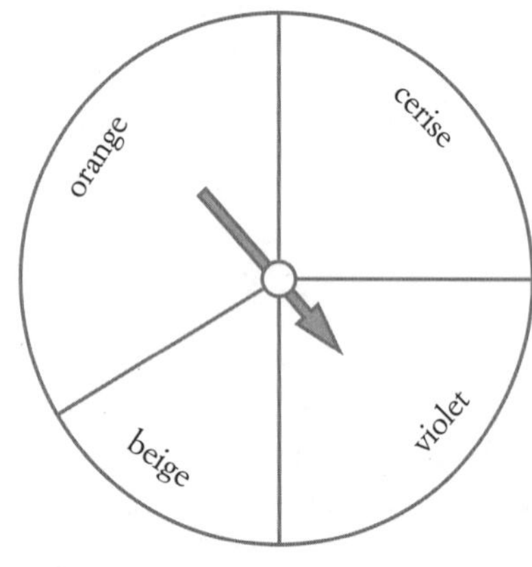

4 From a normal deck of cards one is selected at random. What is the probability that it is:

a a black 4?

b a picture card?

c an even number?

d an Ace or a red card?

5 What is the probability of an event that has a 50–50 chance?

6 A packet of lollies contains 12 red, 15 green, 5 purple and 8 yellow lollies. If one is selected at random, what is the probability that it is:

a red?

b yellow or blue?

c neither green nor red?

7 A sample of matchboxes was taken and the number of matches in each box counted.

No. of matches	47	48	49	50	51	52
Frequency	5	13	26	45	10	1

What is the probability that a box selected at random from this sample contains:

a exactly 50 matches?

b more than 50 matches?

c 48 or fewer matches?

8 The weather report said that there is a 35% chance of rain tomorrow. What is the probability that it will not rain? Express the answer as a simplified fraction.

9 **a** How many different outcomes are possible when a coin is tossed twice?

b What is the probability of a head, then a tail?

c What is the probability of getting a head and a tail in any order?

10 A raffle has 250 tickets and 4 prizes. Li buys one ticket. What is the probability he wins a prize?

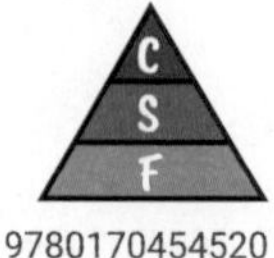

9780170454520

11 What is the probability that a person chosen at random has a birthday in a month beginning with A?

12 Each letter of the word MILLENNIUM is written on a card. If one is drawn randomly, what is the probability that it is:

a M?

b a consonant?

c not L or N?

13 A coin is tossed and a die is rolled.

a How many different outcomes are possible? List them.

b What is the probability of a head and a 5?

c What is the probability of a getting a head and an even number?

14 Two dice are rolled and the difference between the higher and lower numbers is calculated.

a Use this table to calculate all possible differences.

2nd die

1st die		1	2	3	4	5	6
	1				3		
	2		0				
	3						
	4		2				2
	5						
	6						

b What is the probability of rolling a difference of 3?

c Which difference has the lowest probability and what is that probability?

C
S
F

15 A study of 3-child families counted the number of girls in each.

No. of girls	Frequency
0	62
1	144
2	152
3	42

What is the relative frequency (as a percentage) that a family selected at random from this sample has:

a exactly 1 girl?

b more girls than boys?

16 **a** List all possible arrangements of boys and girls for a 3-child family.

b Hence, what is the theoretical probability that a 3-child family has exactly 1 girl? Express the answer as a percentage.

17 A die is biased so that a 6 is twice as likely to occur as the other numbers. What is the probability of rolling an odd number on this die?

WS WORKSHEET

WS WORKSHEET

9 DICE PROBABILITY

YOU'LL NEED A PAIR OF DICE FOR THIS PRACTICAL ACTIVITY.

WORK IN GROUPS OF 2 OR 3

When 2 dice are rolled, there are 36 different possibilities. The sum of the 2 dice can range from 2 to 12.

1 Complete the addition table to list all 36 possible outcomes.

1st die (columns), 2nd die (rows)

+	⚀	⚁	⚂	⚃	⚄	⚅
⚀						
⚁			5			
⚂				7		
⚃				8		
⚄						
⚅						

2 By counting the number of ways of rolling each sum, complete the table.

Sum	Probability	Probability (%)
2	$\frac{1}{36}$	2.78%
3		
4		
5	$\frac{1}{9}$	11.11%
6		
7		
8		
9		
10	$\frac{1}{12}$	8.33%
11		
12		

3 a Which sum has the highest chance? ________

b Which has the lowest chance? ________

4 Note the *symmetry* in the probabilities. Which sums have the second-highest chance? ________

5 Roll a pair of dice 72 times. Write your results in the frequency table below. How were the expected frequencies calculated?

____

____

____

Score	Tally	Frequency	Expected frequency
2			2
3			4
4			6
5			8
6			10
7			12
8			10
9			8
10			6
11			4
12			2

9780170454520

How do your results (frequencies) compare with the expected frequencies?

____

____

____

6 The **law of averages** states that the experimental frequencies should become closer to theoretical (expected) frequencies as more trials (rolls) are taken. Combine results from all groups in your class.

Write them in the table.

Score	Frequency	Expected frequency
2		
3		
4		
5		
6		
7		
8		
9		
10		
11		
12		
Totals		

7 Are experimental frequencies closer to the expected frequencies now? ______________

8 Construct a frequency histogram of these results. Is the histogram symmetrical?

WS WORKSHEET

PS PUZZLE SHEET

9 PROBABILITY CROSSWORD

THIS CROSSWORD WILL REVISE THE LANGUAGE OF CHANCE.

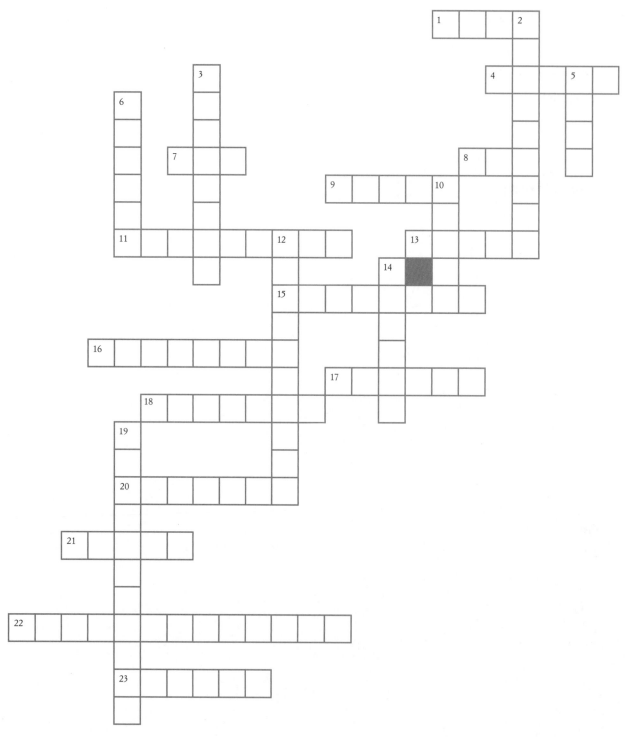

ACROSS

1 The probability that the next baby born is a girl

4 The complement of tossing tails on a coin

7 The number of possible outcomes when a die is rolled

8 The highest probability value

9 The '*E*' in *P*(*E*)

11 Events that have no outcomes in common are called mutually ______________

13 The probability of tossing tails on a coin, as a percentage

15 Another word for likely

16 A red suit in a deck of cards, but not hearts

17 A table that shows the number of items belonging to overlapping categories (2 words)

18 Word to describe something that 'must happen'

20 A result of a situation or experiment, such as tails when a coin is tossed

21 See **6 down**

22 All of the outcomes that are NOT the event

23 The highest sum if 2 dice are rolled

DOWN

2 The number of times an event occurs over repeated trials

3 Describes a low chance

5 In a soccer match, the result could be a win, loss or ______________

6 and 21 across The set of all possible outcomes in a probability situation (2 words)

10 One go or run of a repeated probability experiment

12 Describes an event with zero probability.

14 Describes a situation where every possible outcome is equally likely

19 Mathematical word for chance

9 PROBABILITY 1

Name:

Due date:

Parent's signature:

Part A	/ 8 marks
Part B	/ 8 marks
Part C	/ 8 marks
Part D	/ 8 marks
Total	/ 32 marks

2 COMPLEMENTARY ANGLES ADD UP TO 90°, BUT THE PROBABILITIES OF 2 COMPLEMENTARY EVENTS ADD UP TO 1.

PART A: MENTAL MATHS

Calculators not allowed

1 Complete:

a 3286 mm = ________ m

b 2.4 m^2 = ________ cm^2

2 Decrease \$470 by 10%.

____

3 Find the range of this set of data.

1 2 3 4 5 6

____

4 Simplify: $5a \times (-2b) \times 7b$ ________________

5 Convert $\frac{1}{3}$ to a decimal.

____

6 Evaluate $15.26 \div 0.7$.

____

7 Find f.

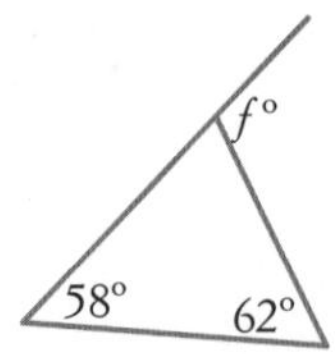

____

PART B: REVIEW

1 Convert $\frac{17}{40}$ to:

a a decimal ________________________

b a percentage ________________________

2 List all possible outcomes for the result of a soccer match between France and South Africa.

____

____

3 Complete: The value of a proper fraction is always between 0 and ________.

4 Match a chance word from the list below to each event shown:

certain likely unlikely impossible

a You win a prize

____

b The sun will rise tomorrow morning

____

c You roll a number greater than 2 on a die

____

d You randomly select a blue sock from a drawer containing all red socks

____

C S F

HW HOMEWORK

9780170454520

PART C: PRACTICE

› Probability
› Complementary events
› Venn diagrams

1 How many possible outcomes are there for:

a rolling a die? ____

b selecting a whole number between 20 and 30

____

2 What is the decimal probability of an event that has an even chance? ____

3 If the probability of rain today is 19%, what is the probability that it won't rain today?

____

4 A bag contains one of each Australian coin. If one is selected at random, what is the probability that it is:

a a silver coin? ____

b between 10c and $1? ____

5 In a music class, students can learn piano or singing. The Venn diagram below shows the number of students in each group. What is the probability that a student chosen randomly from this class:

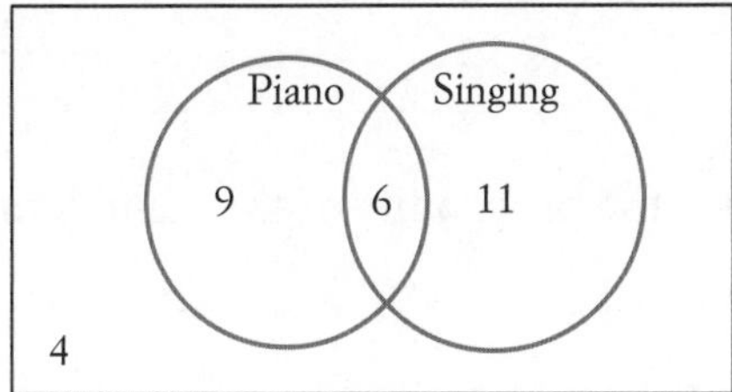

a is in the singing group?

____

b is not in the piano or singing group?

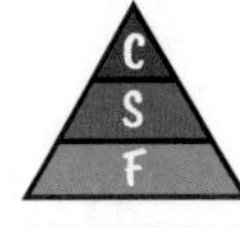

____

PART D: NUMERACY AND LITERACY

1 a List the sample space for the sex of a baby born at a hospital. ____

____

b Explain why the outcomes are mutually exclusive.

____

2 At the party, there were 6 children aged 8, 5 aged 7 and one 6-year-old. If one was randomly selected to start a game, what is the probability of selecting a child aged:

a 7 or more? ____

b more than 7? ____

3 a What is the complementary event to being born in a month beginning with A?

____

____

b What is the probability of this complementary event? ____

4 In a class of 28 students, 14 play tennis and 18 play basketball. There aren't any students who play neither sport.

a Draw a Venn diagram to represent this information.

b What is the probability of randomly selecting from this class a student who plays both tennis and basketball?

____

HW HOMEWORK

9780170454520

9 PROBABILITY 2

Name:

Due date:

Parent's signature:

Part A	/ 8 marks
Part B	/ 8 marks
Part C	/ 8 marks
Part D	/ 8 marks
Total	/ 32 marks

SOME STUDENTS FIND PROBABILITY HARD. IF YOU'RE STRUGGLING WITH THIS, THEN ASK A TEACHER, FRIEND OR PARENT FOR HELP.

PART A: MENTAL MATHS

Calculators not allowed

1 Complete: 3 days = ________ hours

2 Simplify each expression.

a $5a - 4b + 3a - 2b$ ________

b $\dfrac{30ap}{45ap^2}$ ________

3 Find the median of this set of data.

4 Round 162.975 to one decimal place.

____

5 Evaluate $\dfrac{5}{8} \div \dfrac{3}{4}$.

____

6 A plane left Sydney at 13:50 and arrived at Singapore at 22:10 (Sydney time). How long did the plane trip take?

____

7 Find w.

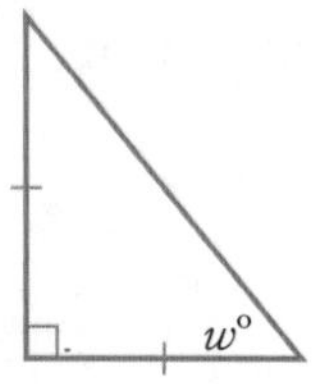

____

PART B: REVIEW

1 Find the probability of:

a tossing heads on a coin ________

b rolling a factor of 4 on a die ________

2 Complete: The sum of the probabilities of an event and its complementary event is equal to ________.

3 There are 5 red, 4 yellow and 6 pink roses blooming in a garden. If I pick a rose at random, what is the probability that it is:

a not yellow? ________

b not white? ________

4 The probability of a train arriving at Burwood station early or on time is 0.6.

a What is the complementary event to a train arriving early or on time?

____

b What is the probability of this complementary event? ________

5 Why is it incorrect to say that the probability of a team winning a netball match is $\frac{1}{3}$ because there are only 3 possible outcomes of a match: win, lose or draw? ________

____

C S F

9780170454520

PART C: PRACTICE

› Two-way tables
› Probability problems
› Experimental probability

1 A group of patients was tested for diabetes and the results are shown below.

	Positive	Negative
Male	45	75
Female	32	68

a How many patients were tested? ____________

b How many males tested negative? ____________

c What is the probability of randomly selecting from this group:

i a male patient?

____

ii a female patient who tested positive?

____

2 a A biased coin was tossed 60 times and tails came up 21 times. What is the relative frequency of throwing a tail on this die?

____

b If this die is tossed 400 times, what is the expected frequency of tails coming up?

____

3 50 students at camp could select kayaking or bushwalking as an activity. Out of these students, 3 became ill and did neither, 9 chose both and a total of 29 students chose kayaking.

a Show this information on the Venn diagram.

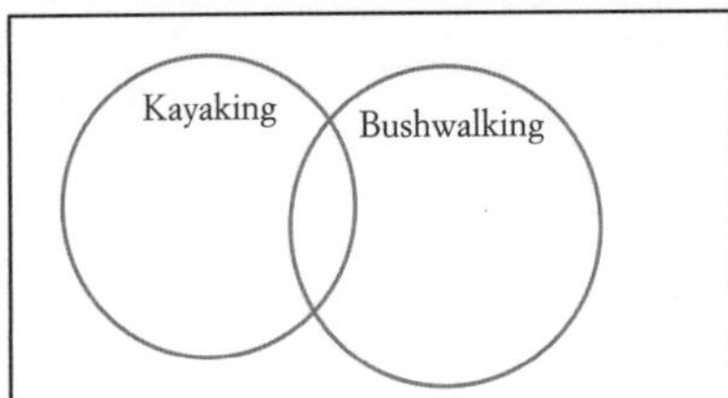

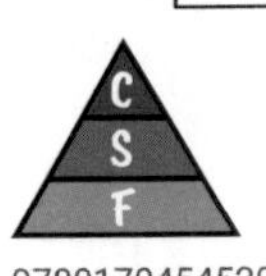

b What is the probability that a student selected randomly chose bushwalking?

____

PART D: NUMERACY AND LITERACY

1 a What is the lowest probability value?

____

b What type of event that has this value?

____

2 As Marshall drives to work there is a 30% chance of a traffic light showing red and a 5% chance of it showing amber.

a Describe the chance of it showing red.

____

b What is the chance that it shows green?

____

c What is the chance that it does not show amber?

____

3 A group of employees were classified according to their age and driving ability.

Age group	Drivers	Non-drivers
Over 40	149	71
40 and under	135	45

a How many non-drivers were there?

____

b Find the decimal probability of randomly selecting from this group:

i a driver aged 40 and under

____

ii a driver or someone aged 40 and under

____

HW HOMEWORK

10 STARTUP ASSIGNMENT 10

BEFORE WE START THE OUR EQUATIONS TOPIC, LET'S REVISE OUR ALGEBRA SKILLS.

WS WORKSHEET

PART A: BASIC SKILLS

/ 15 marks

1 Express each number as a fraction:

a $12\frac{1}{2}\%$

______

b $\frac{7}{12}$ ______

2 Complete: 1 ha = ______ m^2.

3 Test whether this triangle is right-angled.

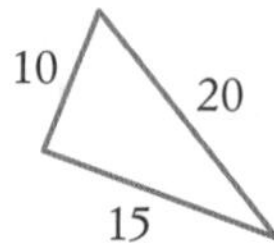

______

4 A group of people were asked how many children they had. The results are shown.

a What was the highest value?

______

b How many people were surveyed?

______

c Calculate to one decimal place the mean.

______

5 If the probability of rain today is 0.35, what is the probability of no rain today?

______

6 What angle is supplementary to 50°?

______

7 Simplify $\frac{2c^2e}{10ce}$. ______

8 Use index notation to simplify $4^6 \div 4^2$.

______

9 Find y.

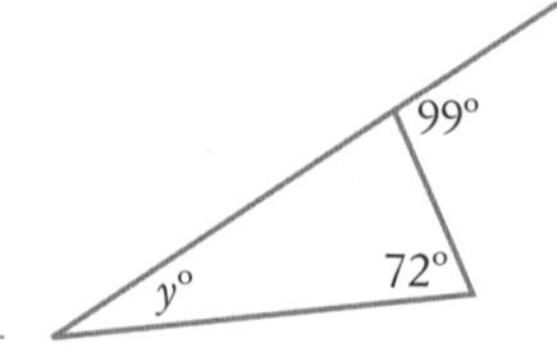

______

10 Express 9 hours as a percentage of one day.

______

11 How many axes of symmetry has an isosceles triangle? ______

12 Calculate to 2 decimal places the area of this quadrant.

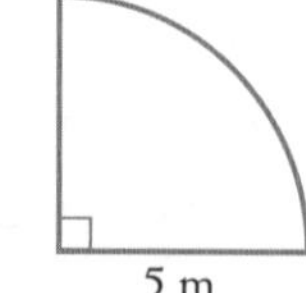

______

PART B: ALGEBRA

/ 25 marks

13 If $z = 6$, then evaluate:

a $2z - 11$ ______

b $\frac{z+9}{3}$ ______

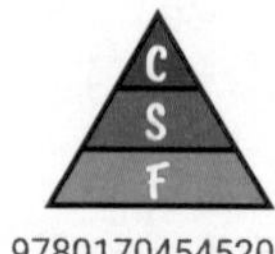

9780170454520

14 Evaluate:

a $3 \times (-2) + 8$

b $-4 \times (-5) - 3$ ______

c $\frac{-6+10}{4}$ ______

15 Simplify:

a $4b + 5 - 6b + 3$ ______

b $3h - 4 + h + 5$ ______

c $m \times 4 + 3$ ______

d $r \times (-2) - 4$ ______

e $u \div 4 + 5$ ______

f $\frac{2d}{9} \times 9$ ______

16 Write an algebraic expression for:

a 3 more than double n

b the difference between 18 and x, where 18 is larger.

17 If $d = -2$, then evaluate:

a $5d + 4$ ______

b $8 - d$ ______

18 Complete:

a $2 \times$ ______ $= -10$

b ______ $- 7 = 6$

c ______ $+ 9 = 5$

d $2 \times$ ______ $+ 5 = 19$

e $5 \times$ ______ $- 8 = 32$

19 If $k = 3$, then show that $\frac{k+7}{5} = 2$.

20 When a number is subtracted from 9 and doubled, the answer is 10. What is the number?

21 Expand:

a $4(n - 7)$ ______

b $5(r + 1)$ ______

c $-2(3d + 9)$ ______

PART C: CHALLENGE

Bonus / 3 marks

James is 12 years older than his sister, Elizabeth. Three years ago, he was twice as old as her.

How old are James and Elizabeth now?

C
S
F

10 WRITING EQUATIONS

SOLVE EACH EQUATION AND
DESCRIBE IN WORDS THE STEPS
TAKEN TO FIND THE SOLUTION.

Equation and solution	Description of steps
1 $5x + 3 = 28$	
2 $3r - 6 = 12$	
3 $\frac{4y}{5} = 8$	
4 $2(n + 7) = 10$	
5 $6a - 6 = 3a + 9$	

9780170454520

I'M ZINA. ALL THE ANSWERS TO THIS CROSSWORD ARE LISTED BELOW. NOW YOU JUST HAVE TO WORK OUT WHERE THEY GO!

EQUATIONS CROSSWORD

PUZZLE SHEET

The answers to this crossword puzzle are listed below in alphabetical order. Arrange them in the correct places.

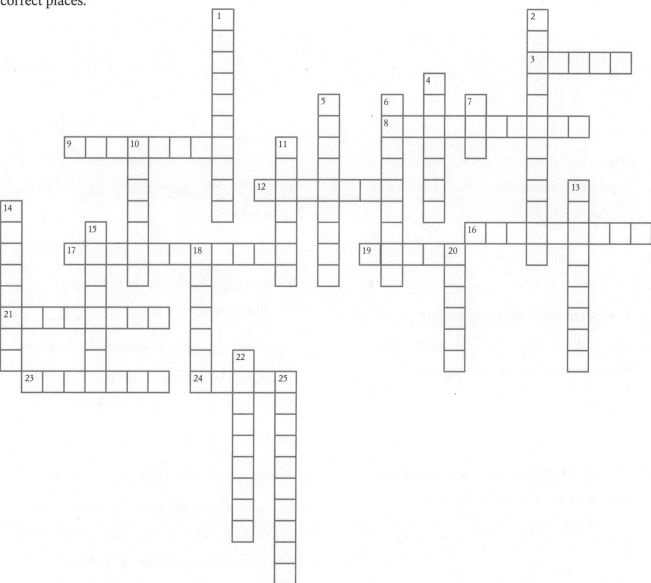

BACKTRACKING	BALANCING	BRACKETS	CHECK
CONSECUTIVE	DIFFERENCE	EQUATION	EXPAND
FORMULA	GUESS	INVERSE	OPERATION
PERIMETER	PRODUCT	PRONUMERAL	REPRESENT
SOLUTION	SOLVE	SUBSTITUTE	SUM
TRANSLATE	TWO-STEP	UNDOING	UNKNOWN
VARIABLE			

10 EQUATION PROBLEMS

THIS WORKSHEET HAS PROBLEMS THAT REQUIRE YOU TO USE EQUATIONS TO SOLVE. DON'T FORGET TO CHECK THAT YOUR ANSWERS WORK.

WS WORKSHEET

Read each problem carefully. Write an equation when necessary, and then solve the problem.

1 7 more than 3 times a number is equal to 16. What is the number?

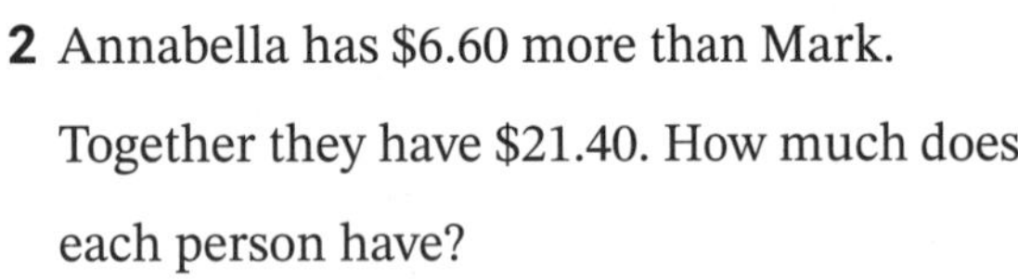

2 Annabella has \$6.60 more than Mark. Together they have \$21.40. How much does each person have?

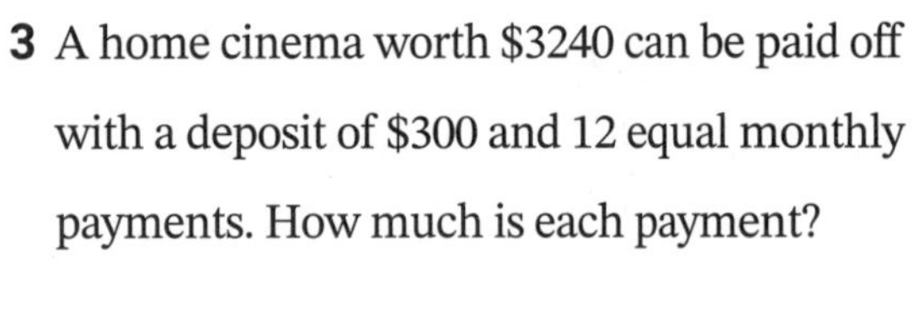

3 A home cinema worth \$3240 can be paid off with a deposit of \$300 and 12 equal monthly payments. How much is each payment?

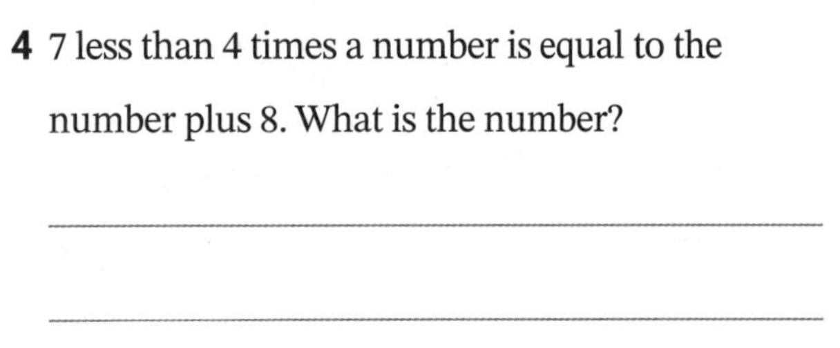

4 7 less than 4 times a number is equal to the number plus 8. What is the number?

5 Daniel is 2 years younger than Christina. Together their ages add to 36. How old is Daniel?

6 The formula for finding the number of toothpicks, t, to build a row of h house shapes is:

$$t = 5h + 1.$$

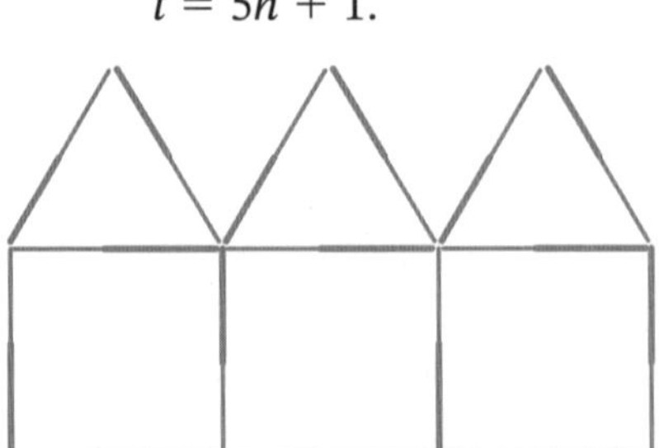

a How many toothpicks are required to build 16 houses?

b How many houses can be built with 56 toothpicks?

7 A rectangle is twice as long as it is wide. If its perimeter is 72 cm, what are its dimensions?

8 In a bowl of fruit, the apple has a mass twice that of the banana, and the mass of the orange is 28 g more than the mass of the banana. If the mass of all 3 fruits together is 124 g, what is the mass of each fruit?

9780170454520

WS WORKSHEET

9 Evan the taxi driver charges fares according to the formula $C = 2.5d + 3.6$, where d is the distance travelled in kilometres and C is the fare in dollars.

a How much does Evan charge for a 28 km trip?

____

b Evan charged $133.60 for a trip. What was the distance travelled?

____

10 The sum of 2 consecutive numbers is 65. What are the 2 numbers?

____

____

11 3 more than a number, all divided by 2, is equal to 5. What is the number?

____

____

12 Jess wants to make a triangular flower bed with 2 sides equal and each twice as long as the third side. She has 48 metres of edging. How long will each side be?

____

____

13 Ronan, Chad and Atif have $116 between them. Ronan has $12 more than Chad, and Atif has $7 less than Ronan. How much does each boy have?

____

____

14 The formula for converting Celsius temperatures (C) to Fahrenheit temperatures (F) is $F = \frac{9C}{5} + 32$.

a Convert 25°C to °F.

____

b Convert 100.4°F to °C.

____

15 The perimeter of this rectangle is 80 cm.

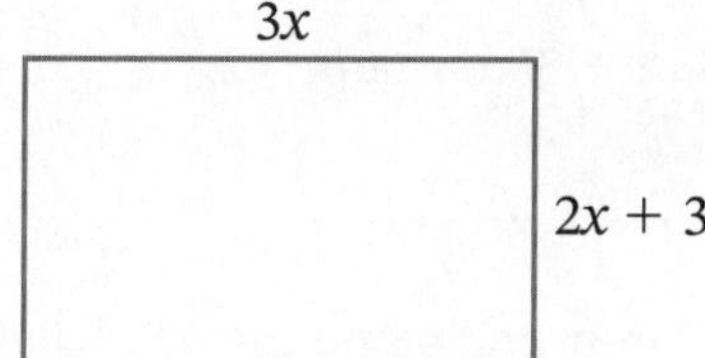

a Find x.

____

b Find the dimensions of this rectangle.

____

10 EQUATIONS 1

MATHS ISN'T EASY, BUT IT'S WORTH DOING. KEEP TRAINING AND WORKING HARD, IT WILL MAKE YOU SMARTER AND FASTER.

Name:

Due date:

Parent's signature:

Part A	/ 8 marks
Part B	/ 8 marks
Part C	/ 8 marks
Part D	/ 8 marks
Total	/ 32 marks

HW HOMEWORK

PART A: MENTAL MATHS

Calculators not allowed

1 Write down the factors of 32.

2 Evaluate $420 \div (20 - 6 \times 3)$.

3 Convert 65% to a decimal.

4 Simplify $4ab - 3ba + 6ab - 2$.

5 (2 marks) Draw and name the quadrilateral that has 2 pairs of equal adjacent sides.

6 A plane leaves Brisbane at 7.55 p.m. and travels for 7 h 32 min. What time does it land?

7 Find the area of this shape.

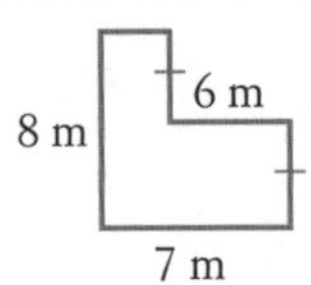

PART B: REVIEW

1 Evaluate each expression.

a $8 \times (-3) + 5$

b $20 - 2 \times (-1)$

2 What is the opposite of:

a adding 4?

b dividing by 5?

c subtracting 6?

3 If $d = -2$, then evaluate each expression.

a $3d - 4$

b $8 - d$

4 Check that $x = 4$ is the solution to $6x + 9 = 33$.

C S F

9780170454520

PART C: PRACTICE

› One-step equations
› Two-step equations

1 Solve each equation.

a $m + 6 = 8$

b $x - 8 = -2$

c $8y = 42$

d $2a - 4 = 8$

e $3m + 5 = 16$

f $5r + 12 = -28$

g $\frac{x}{4} - 3 = 6$

h $\frac{8-a}{3} = -1$

C
S
F

PART D: NUMERACY AND LITERACY

1 What word is used to describe the 'correct answer' to an equation?

2 **a** Write an algebraic expression for the difference between 12 and the product of 4 and n.

b Find n if the difference between 12 and the product of 4 and n is 28.

3 **a** Describe in words the meaning of the equation $\frac{k+9}{2} = 10$.

b Solve $\frac{k+9}{2} = 10$, showing the inverse operations clearly.

c Check that your solution is correct.

4 **a** Write an equation that says that the sum of double m and 8 equals 15.

b Find the value of m.

HW HOMEWORK

10 EQUATIONS 2

Name:

Due date:

Parent's signature:

Part A	/ 8 marks
Part B	/ 8 marks
Part C	/ 8 marks
Part D	/ 8 marks
Total	/ 32 marks

NOTICE THAT THE EQUATIONS ARE GETTING HARDER. SOLVE THESE PROBLEMS, AND YOU WILL MASTER ALGEBRA. GAME ON!

HW HOMEWORK

PART A: MENTAL MATHS

Calculators not allowed

1 Evaluate $\frac{3}{4}-\frac{1}{5}$.

______

2 Complete: 240 g = ______ kg

3 Find w.

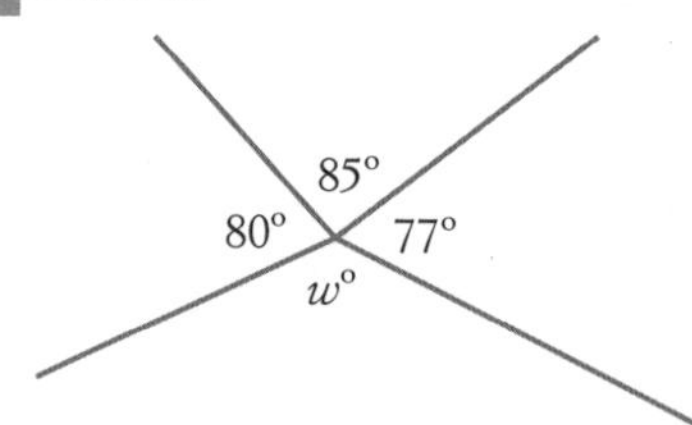

4 Find the percentage probability of randomly selecting a brown chocolate from a box of 12 brown and 8 white chocolates.

______

5 Simplify 28 : 16. ______

6 Find the perimeter of this shape.

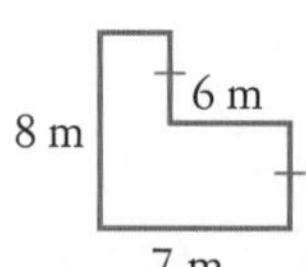

7 Find the lowest common multiple of 8 and 12.

______

8 A prism has a volume of 60 m^3 and a height of 5 cm. Find the area of its base.

______

PART B: REVIEW

1 Expand each expression.

a $4(k - 7)$

______

b $2(5t + 11)$

______

2 Simplify each expression.

a $2m + 8 + 5m - 10$

______

b $7c - 7 - 3c - 9$

______

3 Solve each equation.

a $b + 8 = -3$

______

b $\frac{m}{4} = -2$

______

c $17 - 3w = 29$ ______

______

d $\frac{m}{3} + 2 = 8$ ______

______

C S F

9780170454520

PART C: PRACTICE

› Equations with variables on both sides
› Equations with grouping symbols
› Equation problems

1 (3 marks)

a Solve the equation $2d + 3 = d - 8$.

b Check that your solution is correct.

2 (2 marks) Solve the equation $4(y + 8) = 56$.

3 (3 marks) Katie is 4 years older than her twin sisters. The sum of all their ages is 43.

a Write an equation that describes this situation, if x represents Katie's age.

b Solve the equation to find Katie's age.

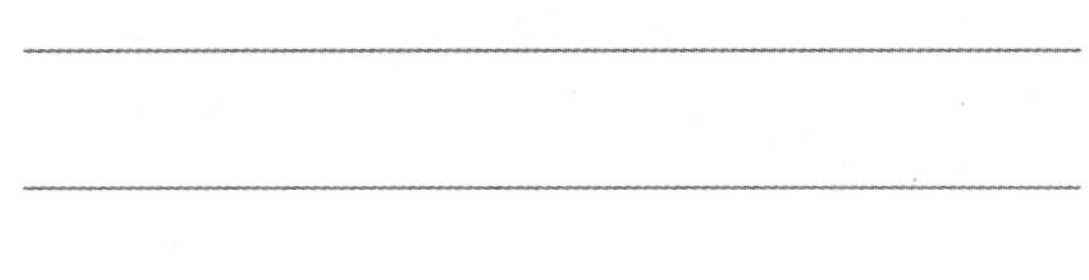

PART D: NUMERACY AND LITERACY

1 Write a one-step equation whose solution is $y = -1$.

2 What is the name given to a letter of the alphabet such as n used to represent a number in an equation? ____________

3 (2 marks) Explain the steps involved in solving the equation $3(2u - 1) = 15$.

4 Write a two-step equation whose solution is $p = 7$.

5 (3 marks) Joe said: 'If you triple my favourite number and add 7 the answer is the same as twice my number plus 10.'

a Solve an equation to find Joe's favourite number.

b Check that your solution agrees with Joe's statement.

10 EQUATIONS REVISION

Name:

Due date:

Parent's signature:

Part A	/ 8 marks
Part B	/ 8 marks
Part C	/ 8 marks
Part D	/ 8 marks
Total	/ 32 marks

WE'RE REACHING THE END OF THE TOPIC, SO I'M GIVING YOU SOME DRILLS OF MIXED QUESTIONS NOW.

PART A: MENTAL MATHS

Calculators not allowed

1 Complete the number pattern:

2, 4, 8, 16, ______

2 Write 6:34 p.m. in 24-hour time. ______

3 Convert $\frac{7}{20}$ to a percentage.

______

4 Find c. ______

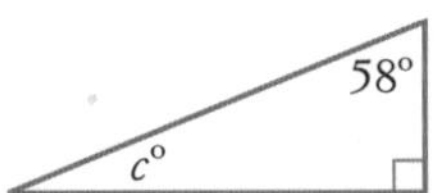

5 Find the decimal probability of randomly selecting a white chocolate from a box of 12 brown and 8 white chocolates?

______

6 Complete: $24 : 40 = 6 :$ ______

7 Factorise $-2d + 14$. ______

8 Find the volume of this square prism.

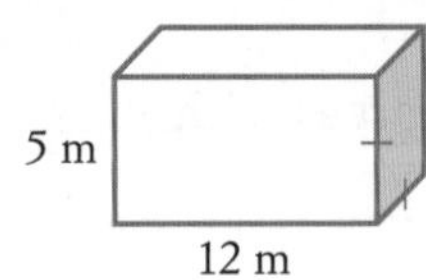

______

PART B: REVIEW

1 Check that $r = -16$ is the solution to $5r + 4 = 3r - 28$. ______

______

2 Check that $u = 6$ is the solution to $6(2u - 5) = 42$. ______

______

3 Solve each equation.

a $w - 6 = -14$

______

b $5b + 8 = 33$

______

c $\frac{m}{4} + 7 = 22$

______

d $\frac{e+3}{2} = 9$

______

e $4w - 18 = 2w - 12$

______

f $5a + 8 = 24 - 3a$

______

C S F

9780170454520

PART C: PRACTICE

› Equations revision

1 (2 marks) Triple a number less 8 is equal to twice the number plus 5. Solve an equation to find the number. ______

______

______

2 Solve each equation.

a $3(x - 4) = -24$

______

______

b $-5(2a + 6) = 20$

______

______

3 (2 marks) Luis is 5 years younger than his twin brothers. If the sum of all their ages is 124, solve an equation to find Luis' age. ______

______

______

4 A technician charges for fixing washing machines using the formula $C = 32h + 45$, where $\$C$ is the charge and h is the number of hours the job takes.

a Calculate the cost of a 3-hour job.

______

b Find the number of hours worked if the charge is $125.

______

______

C S F

PART D: NUMERACY AND LITERACY

1 List the next 3 consecutive numbers after:

a 15 ______

b y ______

2 The area of this rectangle is 35 m². Find:

$(3x - 2)$ cm

5 cm

a the value of x

______

______

b the perimeter of this rectangle

______

3 3 consecutive numbers have a sum of 156.

a Write an equation that describes this situation.

______

b Solve the equation to find the 3 numbers.

______

______

______

4 A temperature in degrees Celsius (°C) can be converted to degrees Fahrenheit (°F) using the formula $F = \frac{9C}{5} + 32$.

a Convert 30°C to °F.

______

b Convert 50°F to °C.

______

______

HW HOMEWORK

WS WORKSHEET

11 STARTUP ASSIGNMENT 11

QUESTION 2'S ANSWER IS NOT 'DIAMETER' BECAUSE IT DOESN'T GO THROUGH THE CENTRE OF THE CIRCLE.

PART A: BASIC SKILLS

/ 15 marks

1 **a** Write an algebraic expression for the difference between 18 and triple p.

______

b If the difference between 18 and triple p is 30, find p. ______

______

2 Name the part of the circle drawn.

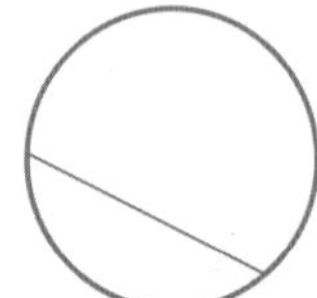

______

3 Find the median of these values:

8, 1, 7, 9, 5, 8, 9, 1, 9.

______

4 Evaluate:

a \$7.00 − \$4.28 ______

b $(-2)^5$ ______

5 Factorise $44m - 8m^2$. ______

6 Test whether this triangle is right-angled.

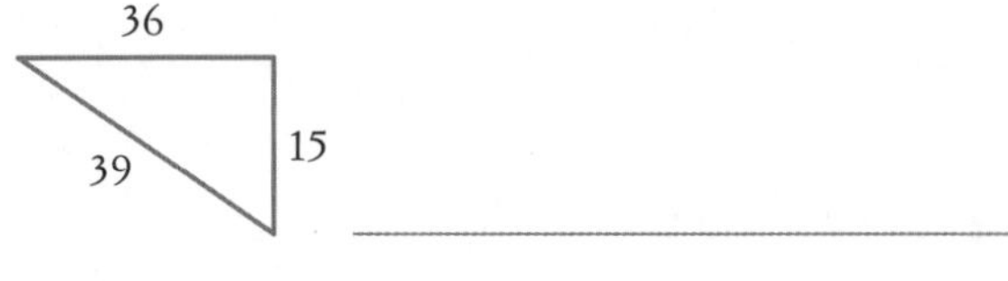

______

______

7 Increase \$40 by 35%.

______

8 Simplify $\frac{6d^2}{3de}$. ______

9 Find to 2 decimal places the area of this semi-circle.

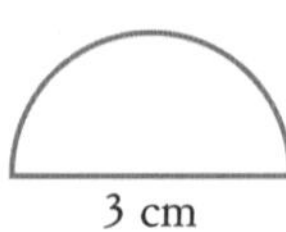

______

______

10 **a** What is the probability of selecting a spades card from a deck of cards?

______

b What is the complementary event to selecting a spades card? ______

______

11 Classify this triangle:

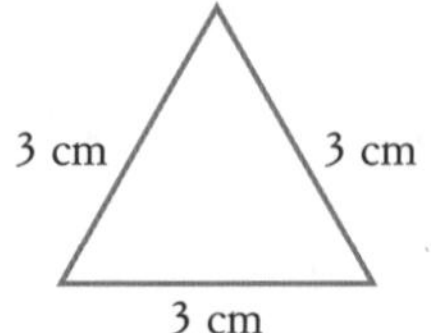

a by sides ______

b by angles ______

PART B: FRACTIONS AND TIME

/ 25 marks

12 Simplify:

a $\frac{20}{45}$ ______

b $\frac{18}{8}$ ______

c 12 : 30 ______

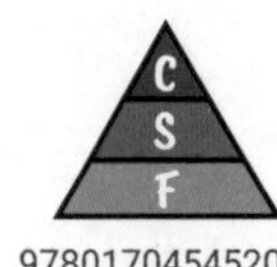

9780170454520

13 Find the highest common factor of:

a 21 and 7 ____________

b 32 and 48 ____________

14 Evaluate 3.5×100. ____________

15 Convert:

a 3 hours and 20 minutes to hours

____

b 360 minutes to hours

____

c 2400 seconds to minutes

____

d 5.2 hour to minutes

____

16 Complete:

a 2.9 L = ________ mL

b 400 g = ________ kg

c 8.51 cm = ________ mm

d 7.19 kL = ________ L

17 Complete:

a $\frac{4}{5} = \frac{20}{}$

b 5 : 6 = ________ : 48

18 Write 7:30 p.m. in 24-hour time.

____

19 Write 18:45 in 12-hour time.

____

20 Find:

a $\frac{2}{5} \times 45$ ____________

b $\frac{3}{10} \times \$150$ ____________

21 A drawer has 4 blue shirts, 11 black shirts and 13 white shirts. What fraction is blue?

____

22 Find the lowest common multiple of

a 10 and 6 ____________

b 15 and 5 ____________

23 How many hours between 6 a.m. and 1 p.m.?

____

24 What is the time 8 hours 10 minutes after 4:30 p.m.? ____________

PART C: CHALLENGE

Bonus / 3 marks

This is the famous *Tower of Hanoi* puzzle.

4 discs or coins are stacked at A, in order of size with the smallest at the top. Move the discs one at a time so that they are identically stacked at C.

Discs may be placed temporarily at B but a larger disc cannot rest on a smaller one anywhere.

Can you solve the puzzle in 15 moves?

A B C

C S F

9780170454520

WS WORKSHEET

WS WORKSHEET

11 RATIO APPLICATIONS

HERE ARE SOME PROBLEMS INVOLVING RATIOS, WHICH COMPARE PARTS OR SHARES OF THINGS.

1 Two lengths of wood are cut from one plank in the ratio 4 : 7. If the longer piece measures 35 cm, what is the length of the shorter piece?

____

____

2 The directors of Centuryworld share its profits each year in the ratio 1 : 2 : 3. The managing director, who receives the greatest share, received $9 144 000 last year. What was the Centuryworld profit last year?

____

____

3 A factory produces fuses for cars. On average, 2 fuses in 65 are found to be faulty. In a batch of 325 fuses, how many would you expect to be faulty?

____

____

4 On a student excursion, one teacher is required for every 18 students. If 162 students plan to go on the excursion, how many teachers need to go?

____

5 The lengths of the sides of a triangle are in the ratio 3 : 4 : 5. The perimeter of the triangle is 204 cm. What is the length of each side?

____

6 The masses of 2 packets of detergent are in the ratio 3 : 10.

a If the smaller packet has a mass of 1.5 kg, what is the mass of the larger packet?

____

b If the larger packet costs $12.50 and the smaller packet costs $3.90, which packet is the cheaper per kilogram? By how much?

____

7 Enzo's Produce Store buys fruit and vegetables in the ratio 5 : 7. The mass of the fruit is 8.5 tonnes. What is the total mass of produce ordered?

____

8 The heights of Simon and Joshua are in the ratio 8 : 9. If Joshua is 1.35 m tall, how tall is Simon?

____

9 The ratio of length to width in a rectangle is 7 : 5. The rectangle's perimeter is 168 cm. What are the dimensions of the rectangle?

____

10 In order to make a certain glue, 2 tubes, A and B, must be mixed in the ratio 3 : 1. If we are using 15 mL of A:

a how much of B should we use?

____

b how much glue will be made altogether?

____

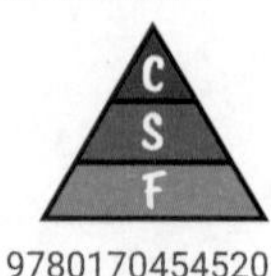

9780170454520

11 Gaz, Judy and Kevin share a lottery prize in the ratio 2 : 3 : 1. How much does each receive if the prize was $150 000?

____

____

12 A rope is cut into 2 pieces in the ratio 3 : 4. The longer piece is 116.2 m. What was the original length of the rope?

____

____

13 When making concrete, sand and cement are mixed in the ratio 4 : 1. If 150 kg of concrete is required, how much sand and how much cement is needed?

____

____

14 A 9.5 m length of timber is used to make a rectangular frame with dimensions in the ratio 3 : 2. Find the length and width of the frame.

____

____

15 Chemicals X, Y and Z are to be mixed in the ratio 3 : 7: 10. If 45 mL of X is used, how much of each of the others will be used?

____

____

16 The prices of LPG and petrol are in the ratio 3 : 8.

a If LPG is 51.3 cents per litre, what is the price of petrol?

____

____

b If LPG goes up by 3 cents per litre, how much does the price of petrol rise to keep the prices in the same ratio?

____

____

WS WORKSHEET

C S F

11 TRAVEL GRAPH STORIES

TRAVEL GRAPHS ARE ALSO CALLED 'DISTANCE-TIME GRAPHS'. CAN YOU GUESS WHY?

Choose a story and use the grid below to draw a travel graph to represent the trip taken.

LARA'S STORY: Lara went to visit her friend, Courtney. She left home at 9 a.m. and, after travelling for 45 minutes, stopped at a car wash 30 km from home. After the 15-minute wash, she continued travelling at an average speed of 60 km/h until she reached Courtney's house $1\frac{1}{2}$ hours later. Lara stayed at Courtney's for $3\frac{1}{2}$ hours before starting her journey home. The first 80 km was completed in one hour, but then Lara was stuck in roadworks at a standstill for 30 minutes. Afterwards, she continued home at a constant speed, arriving at 5 p.m.

HAYDEN'S STORY: Hayden works for Carlon Consulting, 55 km from home. He left home at 9 a.m. and drove at an average speed of 55 km/h to get to work. At midday, he had lunch at a restaurant 15 km further from home, a 15-minute trip away. After an hour's lunch, Hayden drove at 50 km/h for half an hour towards home to visit a client's office, where he worked for two hours. Afterwards, Hayden went to watch his daughter play soccer at a field that was 5 km from home. The soccer match lasted from 4:30 p.m. to 5:30 p.m. They both went home afterwards.

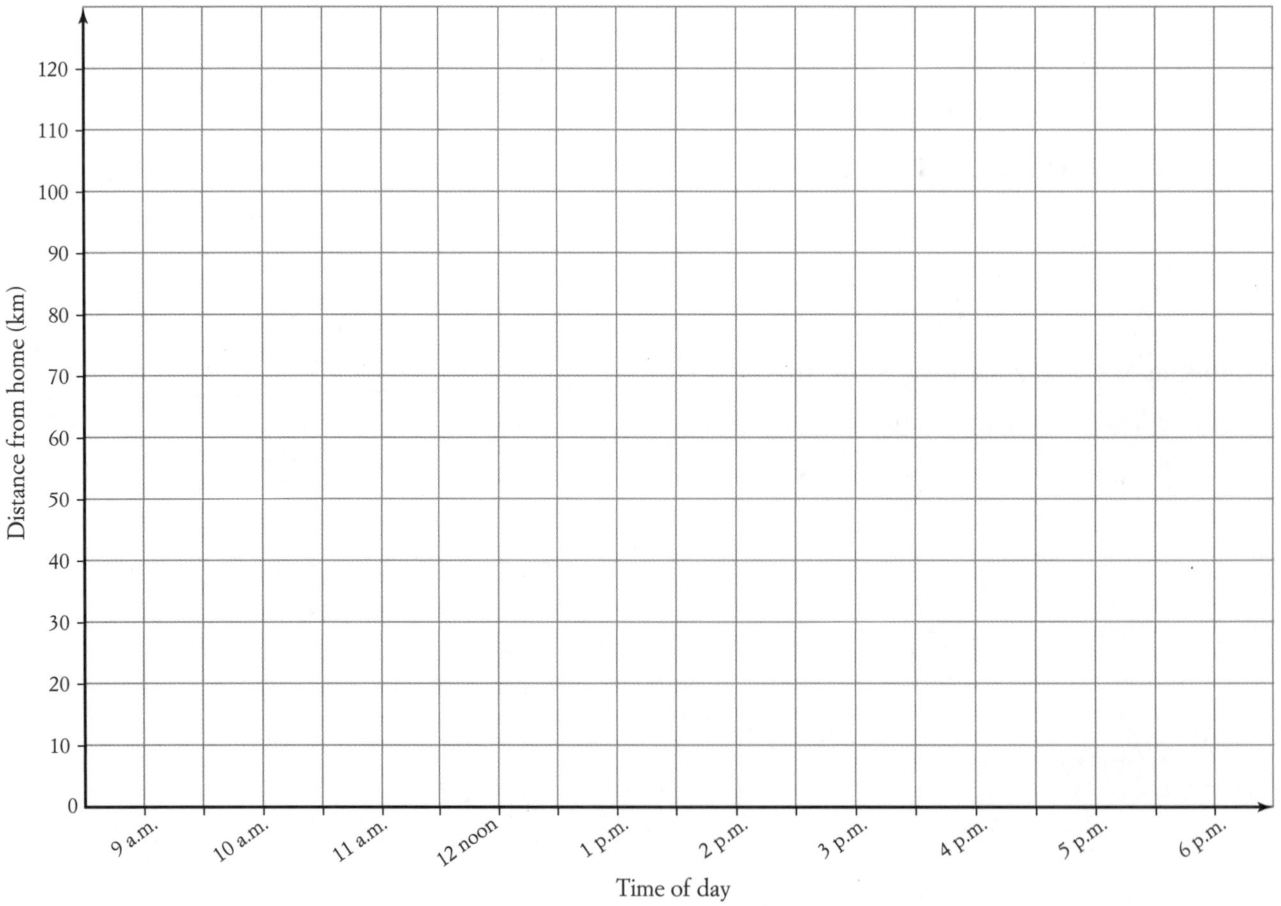

Distance from home (km): 0, 10, 20, 30, 40, 50, 60, 70, 80, 90, 100, 110, 120

Time of day: 9 a.m., 10 a.m., 11 a.m., 12 noon, 1 p.m., 2 p.m., 3 p.m., 4 p.m., 5 p.m., 6 p.m.

9780170454520

RATIOS, RATES AND TIME FIND-A-WORD

PUZZLE SHEET

P N N K T O S G M Z J G L P J B Y O K L H Q M E L
U K T C I E E P Q D B X G D S G I R A P T M T V V
N V A J A Y R S E E D L W Q W I G M A Q O A I D M
I T C E L G K M H E W U K Q N F R R Z E R O A E N
T C J Q E M Z I D T D L E R P O G O P K C Y T E O
A P Z B P I N E A O A Q Y Q F C N Y L A L R C W I
R D Q F T D E W E A H Z F N O E G W E I M H W K T
Y U B J N W L E H S M D I S M A J N G R A T I O C
L C W W X Y S R A E I J T E V M U H I X G J B I A
E C N A T S I D H H C M Y M Q B T I B D B R M Z R
D E L A C S T N Y X E T P H D U R E A L I T Q R F
Y M E F A E I N I W Z E B L M V I A A R L V U E T
L J E T A D M O H P T X X S I E E V Q D E A I P S
I S R I W C E Q I J M A U T N F Q B A C B K X D L
T I M C E I X W A M V H E A A F Y E I L Z F G S M
W D L C R W J R I Y D L G T L B N S J I E H T B A
Z K K U U Q U A N T I T Y I Z E T T L W P N K F N
B F R U R K D O H V D X L O G M V K U B A C T T C
H L Q X U N S Y C Q R K R N Z L B A L U W H F M I
Q L G U H P T T D D G D S A J S L V R C M H Z V Z
B J Z R U B O K N G X W S R E A W M V T F A F K M
I N T E R N A T I O N A L Y I N Z F W N Y D U E H
I X E K F X K K D M X O I M N B I U J A D P J I M
V V L V G Q A I F M M F I M D Q X L G L O J Z P Y
C G G T S N S W A B V N M S Z Y R U C P C G K M K

Find these words in the puzzle above. They are across, up and down, and diagonal, and can be backwards as well as forwards.

AHEAD BEHIND BEST BUY
COST DATE DAYLIGHT DISTANCE
DIVIDING EQUIVALENT FRACTION GRAPH
INFORMAL INTERNATIONAL LINE MAP
PER PLAN QUANTITY RATE
RATIO REAL SCALED SIMPLIFY
SPEED STATIONARY TERM TIME
TRAVEL UNITARY ZONE

11 RATIOS 1

LET'S REVISE FRACTIONS AND RATIOS.

Name:

Due date:

Parent's signature:

Part A	/ 8 marks
Part B	/ 8 marks
Part C	/ 8 marks
Part D	/ 8 marks
Total	/ 32 marks

PART A: MENTAL MATHS

Calculators not allowed

1 Evaluate $\sqrt{5\times5\times5\times5}$. ______

2 Convert 12% to:

a a simple fraction ______

b a decimal ______

3 Solve the equation $6a + 3 = 42$.

______

4 Find the area of this triangle.

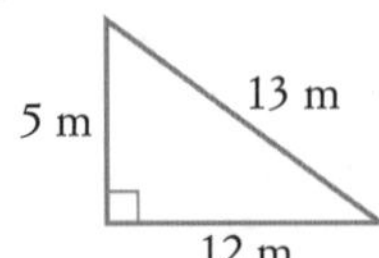

5 Simplify: $12a \times (-5b)$. ______

6 Draw a parallelogram.

7 Calculate the mean of this set of data.

1 2 3 4 5 6

PART B: REVIEW

1 Simplify each fraction.

a $\frac{8}{20}$ ______

b $\frac{45}{15}$ ______

2 Find the lowest common multiple of 9 and 12.

______

3 Find $\frac{3}{8}$ of 1 day (in hours).

______

4 Complete:

a $\frac{5}{7} = \frac{}{28}$

b 1 hour = ______ seconds

c 4.4 kg = ______ g

5 At the cinema, 45% of the audience were children. If there were 144 children, how many people were there in the audience?

______

______

______

C S F

9780170454520

PART C: PRACTICE

› Equivalent ratios
› Simplifying ratios
› Ratio problems

1 Complete:

a 3 : 5 = 12 : ______

b 48 : 30 = ______ : 5.

2 To make concrete, a builder mixes sand and cement in the ratio 5 : 4. If a mix of concrete contains 25 kg of sand, find the amount of cement. ______

3 Simplify each ratio.

a 39 : 21 ______

b 0.25 : 7.5 ______

c $\frac{2}{5} : \frac{3}{4}$ ______

d 40 cm : 5 m ______

4 Grace and Liong share the rent in the ratio 7 : 5. If Liong pays $165, find the total rent.

______

PART D: NUMERACY AND LITERACY

1 For this shape, write as a simple ratio:

a shaded parts to unshaded parts

______

b unshaded parts to the whole shape.

______

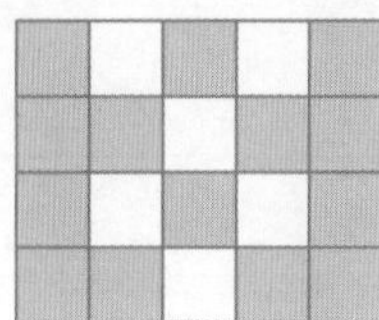

2 The annual birth rate (births : population) in Australia is approximately 12 : 1000.

a Explain in words what this means.

______

b If Australia's population this year is 25 000 000, approximately how many babies will be born this year?

______

3 Complete:

a To simplifying a ratio involving whole numbers, divide all terms by the highest ______ ______.

b To simplifying a ratio involving decimals, first multiply all terms by a ______ of 10 to make them whole numbers.

4 The ratio of staff to students at Westvale High School is 2 : 45.

a If there are 42 teachers, how many students are there?

______

b If next year there are 990 students at the school, how many teachers will there be?

______

HW HOMEWORK

11 RATIOS 2

Name:

Due date:

Parent's signature:

THIS ASSIGNMENT HAS RATIO PROBLEMS FOR YOU TO PRACTISE ON. CAN YOU SOLVE THEM ALL?

Part A	/ 8 marks
Part B	/ 8 marks
Part C	/ 8 marks
Part D	/ 8 marks
Total	/ 32 marks

PART A: MENTAL MATHS

Calculators not allowed

1 (2 marks) Mark a pair of alternate angles on parallel lines and write the rule about them.

2 Evaluate $\sqrt[3]{4 \times 4 \times 4}$ __________

3 Convert 12.5% to a simple fraction. __________

4 Solve this equation $\frac{5a}{8} = 4$.

5 Find the area of this parallelogram.

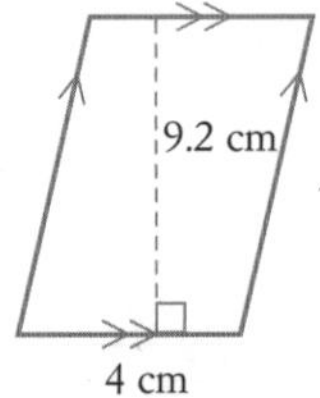

6 Find the value of $4a - 5$ if $a = 6$. __________

7 A parallelogram has one angle 68°.

Find the sizes of the other 3 angles.

____

8 Find the mode of this set of data.

1 2 3 4 5 6

____

PART B: REVIEW

1 Complete:

a 20 : 12 = 5 : __________

b 5 : 7 = __________ : 63

2 The ratios of boys to girls in Year 8 is 5 : 6.

If there are 78 girls in Year 8, how many:

a boys are there?

____

b students in Year 8 altogether?

____

3 Simplify each ratio.

a 15 : 35 __________

b 42 : 14 __________

c 0.4 : 5.2 __________

d 5 min : 1.5 h __________

C S F

9780170454520

PART C: PRACTICE

› Ratio problems
› Scale maps and plans
› Dividing a quantity in a given ratio

1 To make concrete, you mix cement, gravel and sand in the ratio 1 : 2 : 4. If Goran uses 20 kg of gravel, find the mass of:

a the cement

______

b the total concrete mix

______

2 Google Maps use a scale of 3 cm : 2 km.

a Simplify this ratio.

______

b The scaled distance between 2 locations on the map is 16.5 cm. Find the actual distance.

______

3 Divide $720 in the ratio 5 : 3.

______

______

4 Chakrika's soccer team won 7 matches for every 4 matches it lost. If they played 44 matches this season, how many games did they lose?

______

5 In a town of 45 000, the ratio of left-handed people to right-handed people is 2 : 13. Calculate:

a the number of left-handed people

______

b how many more right-handed people there are than left-handed people

______

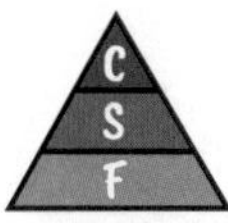

PART D: NUMERACY AND LITERACY

1 Explain what it means when a scale plan or drawing has a scale of:

a 1 : 500 ______

______

b 25 : 1 ______

______

2 When dividing a quantity in a ratio of 8 : 3, what is the first step?

______

3 (4 marks) Nadine, Janine and Renee shared a Lotto entry, paying $20, $15 and $5 respectively.

a What is the ratio of their shares of their entry in simplest form?

______

b They won a prize of $56 000 to divide in the ratio of their shares. Calculate how much of the prize each person should receive.

______

______

______

c What is one way of checking that your answer to part **b** is correct?

______

4 Jackson mixes cordial and water in the ratio 1 : 6 to make a drink, while Stefan mixes cordial and water in the ratio 1 : 8. Explain whose drink is sweeter.

______

______

HW HOMEWORK

11 RATES

Name:

Due date:

Parent's signature:

Part A	/ 8 marks
Part B	/ 8 marks
Part C	/ 8 marks
Part D	/ 8 marks
Total	/ 32 marks

RATES COMPARE 2 MEASUREMENTS WITH DIFFERENT UNITS, SUCH AS KM/H FOR SPEED. SOLVE THESE PROBLEMS INVOLVING RATES.

HW HOMEWORK

PART A: MENTAL MATHS

Calculators not allowed

1 Evaluate each expression.

a $\sqrt[3]{3\times3\times3\times5\times5\times5}$ ______

b 5.8×0.02

______

2 Solve the equation $2a - 8 = 5$.

______

______

3 Find the circumference of this circle using $\pi = 3.14$.

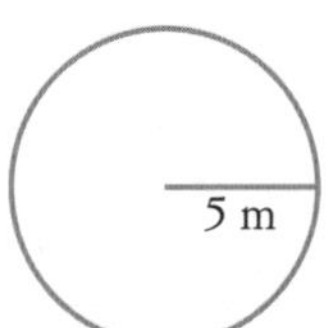

______

4 Convert 32% to a decimal. ______

5 Find the value of $3m + 2$ if $m = -4$. ______

6 Name the quadrilateral with 4 equal sides.

______

7 Complete: Complementary angles

______

PART B: REVIEW

1 Convert $16.50 into cents. ______

2 Simplify each rate.

a 300 words in 5 minutes

______

b 440 km in 4 hours

______

c $85 for 5 hours' work

______

3 Complete:

a 8.1 L = ______ mL

b 3.25 m = ______ mm

4 Jordan buys 8 ice creams for $28.40. What is the cost of 5 ice creams?

______

______

5 At Westvale High School, 272 of students speak a second language. If this represents 32% of the student population, how many students are there in the school?

______

______

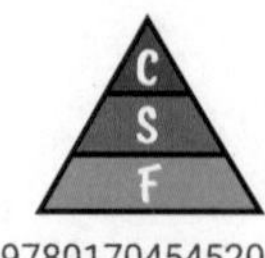

9780170454520

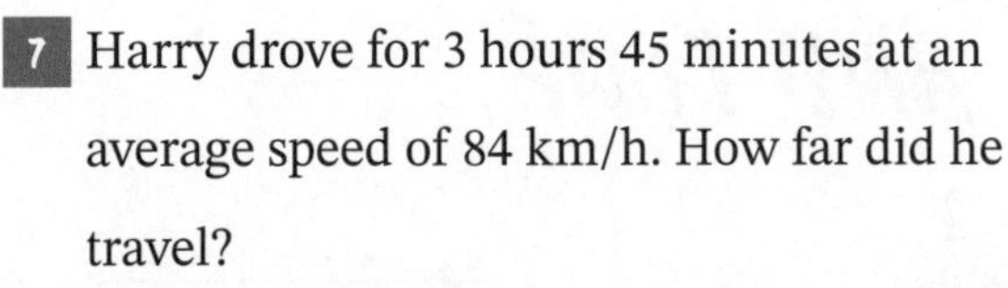

PART C: PRACTICE

› Best buys
› Rate problems
› Speed

1 Hannah drove 736 km in 8 hours. Calculate her average speed.

2 Ziad has a typing speed of 86 words/min.

a How many words can he type in 15 minutes?

b How long will it take him to type up a 2000-word report? Answer to the nearest minute.

3 Which size of rice is the best buy?

A \$4.70 for 2 kg

B \$6.90 for 3 kg

C \$10.50 for 5 kg

4 If petrol costs 149.2 c/L, calculate the cost of 45 L of petrol.

5 How long will it take a truck to travel 540 km if its average speed is 92 km/h? Answer in hours and minutes.

6 Stephanie earns \$16.10 per hour. If she earned \$515.20 last week, how many hours did she work?

7 Harry drove for 3 hours 45 minutes at an average speed of 84 km/h. How far did he travel?

PART D: NUMERACY AND LITERACY

1 Write the formula for average speed.

2 **a** Michael drove for 520 km at an average speed of 80 km/h. How long did the journey take?

b If he left home at 8 a.m. what time did he arrive if he stopped for lunch for 45 minutes?

3 **a** Write 'km/h' completely in words.

b What does it mean to say that a car is travelling at a speed of 85 km/h?

4 Complete: To find the best buy, compare the ________ price of each item and choose the item whose price is the ________.

5 Jake scored 18 goals in 8 soccer matches while Zac scored 25 goals in 10 matches.

a What was Jake's goal rate per match?

b Who had the better rate of goal scoring?

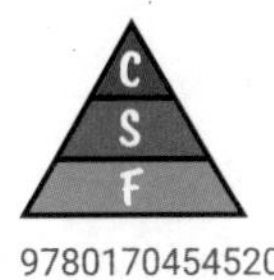

11 TRAVEL GRAPHS AND TIME

WHAT'S THE FORMULA FOR SPEED AGAIN?

Name:

Due date:

Parent's signature:

Part A	/ 8 marks
Part B	/ 8 marks
Part C	/ 8 marks
Part D	/ 8 marks
Total	/ 32 marks

HW HOMEWORK

PART A: MENTAL MATHS

Calculators not allowed

1 (2 marks) Mark a pair of co-interior angles on parallel lines and write the rule about them.

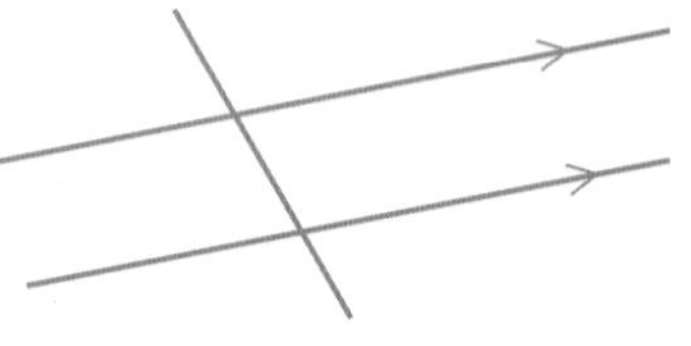

2 Convert $\frac{5}{8}$ to a decimal.

3 Evaluate each expression.

a 7^3

b $4.2 - 0.3 \times 0.5$

4 Find the area of this circle using $\pi = 3.14$.

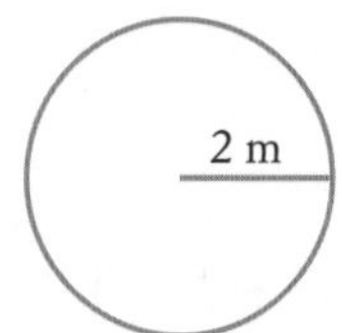
2 m

5 Use index notation to simplify $5^6 \div 5^3$.

6 Expand and simplify $7(r + 6) - 3(2r - 4)$.

PART B: REVIEW

1 Write 4:25 p.m. in 24-hour time.

2 Write 19:50 in 12-hour time.

3 Complete: 100 min = ______ h ______ min.

4 How many hours between 9 a.m. and 5 p.m.?

5 Chris took 4.5 hours to drive 405 km.

a What was his average speed?

b How long will it take him to travel 300 km at this speed? Answer in hours and minutes.

6 What is the time 4 hours and 35 minutes after 7:20 a.m.?

7 Which brand of jam is the best value for money?

A Jim Jam: $1.85 for 150 g

B Jam I Am: $2.45 for 200 g

C T'rific Jam: $3.30 for 275 g

C S F

9780170454520

PART C: PRACTICE

› Travel graphs
› Timetables and time calculations

1 This travel graph shows Grace's journey.

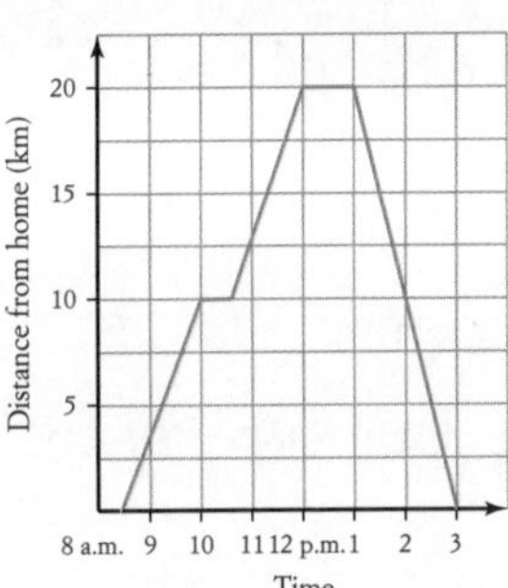

a When did Grace leave home? ________

b What happened at 10 a.m.?

____

c When was Grace furthest from home?

____

d How long did she stop before returning home?

____

e What was her speed while returning home?

____

2 The is part of a train timetable.

Yenley	9.25	9.45	10.05
Tisdell	9.48	10.06	10.28
Bobtown	9.55	10.11	10.35
Clifford	10.07	10.21	10.47
Sandilands	10.21	10.35	11.01

a How long does the 9.48 train take to go from Tisdell to Clifford?

____

b If I catch the 10.05 train at Yenley, where will I be after 30 minutes?

____

c If I need to be in Sandilands by 10.45, what is the latest train I can catch from Bobtown?

____

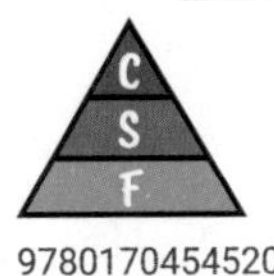

PART D: NUMERACY AND LITERACY

1 (2 marks) Explain how to convert 20:35 to 12-hour time.

____

____

____

____

2 George drove for 9.5 hours at an average speed of 88 km/h.

a How far did he travel?

____

b If he left home at 8:30 a.m., at what time did he arrive if he stopped for lunch for one hour and dinner for 45 minutes?

____

3 On a travel graph, explain what it means when the line:

a becomes flat (horizontal)

____

b points down towards the Time axis

____

4 (2 marks) Jabira invested \$50 000 into her new business and Lily helped her with \$30 000. At the end of the first year they decided to divide their profit of \$29 760 in the same ratio as they had invested.

Calculate how much each person should receive.

____

____

____

____

HW HOMEWORK

12 STARTUP ASSIGNMENT 12

LET'S PREPARE FOR THE GRAPHING LINEAR EQUATIONS TOPIC BY REVISING SOME ALGEBRA AND THE NUMBER PLANE.

WS WORKSHEET

PART A: BASIC SKILLS / 15 marks

1 Evaluate \$50 − \$27.65. ______

2 Write 0.45 as a simple fraction. ______

3 Complete: 1 m^2 = ______ cm^2.

4 Draw a triangular pyramid.

5 A movie starts at 4.30 p.m. and runs for 108 minutes. At what time does it finish?

______

6 What are supplementary angles?

______

______

7 Simplify $\frac{3}{8} \times \frac{2}{15}$.

______

8 Find the lowest common multiple of 15 and 30.

______

9 Evaluate $(-3)^4$. ______

10 Find x.

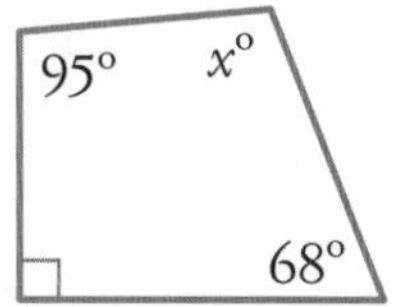

11 Simplify $-6d \times (-5de)$. ______

12 What is the size of one unit on this scale?

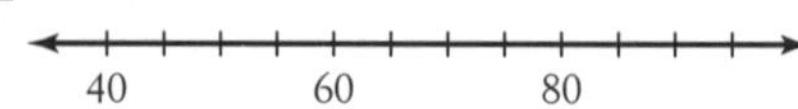

______

13 Expand and simplify $3(p + 5) + 4p - 9$.

______

______

14 Decrease \$260 by 18%. ______

15 What is a scalene triangle? ______

______

PART B: ALGEBRA AND THE NUMBER PLANE / 25 marks

16 For this number plane, write the coordinates of:

a B ______

b C ______

c D ______

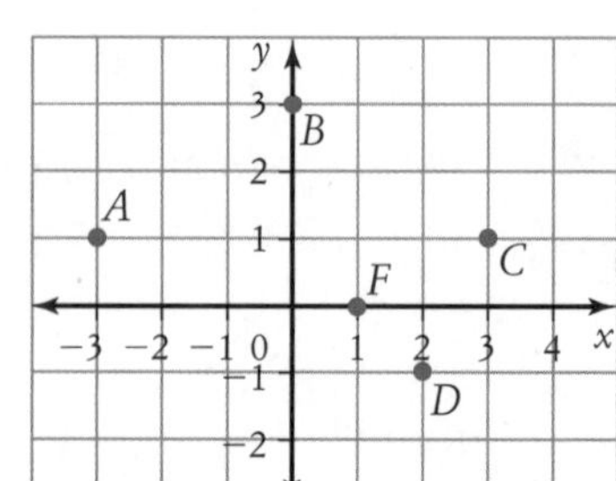

17 For the number plane above:

a plot the point $E(-1, -2)$ ______

b in which quadrant is point A? ______

c which point is in the 4th quadrant? ______

d which point is on the y-axis? ______

9780170454520

18 If $x = 5$, then evaluate:

a $x - 8$ ______

b $3x + 7$ ______

c $-x - 2$ ______

d $5x + 2$ ______

19 Complete this pattern: 4, 7, 10, 13, ______

20 Complete:

a $2 \times 4 +$ ____ $= 10$

b $5 \times 3 -$ ____ $= 7$

21 On a number plane, what are the coordinates of the origin? ______

22 Solve each equation.

a $2x + 9 = 17$

______

______

b $3x - 1 = 26$

______

______

23 This diagram shows a pattern of table and chairs.

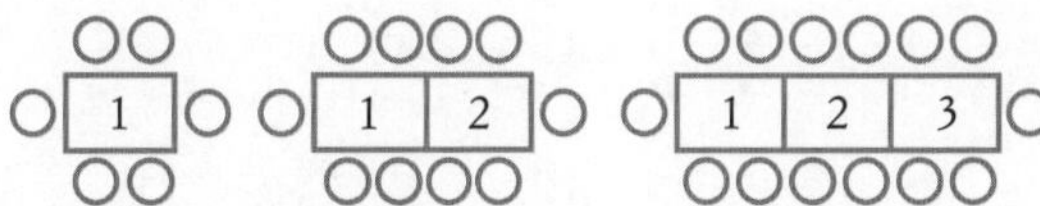

a Complete this table of values for the pattern.

No. of tables	1	2	3	4
No. of chairs	6			

b How many chairs can be seated around 8 tables?

______

______

24 If $y = -2$, then evaluate:

a $y + 4$ ______

b $2y + 7$ ______

c $-3y - 1$ ______

25 (3 marks)

a Which 2 of these points lie on the line below?

$(3, -3), (-3, -3), (2, -1), (1, 1), (2, 7), (-2, 1)$

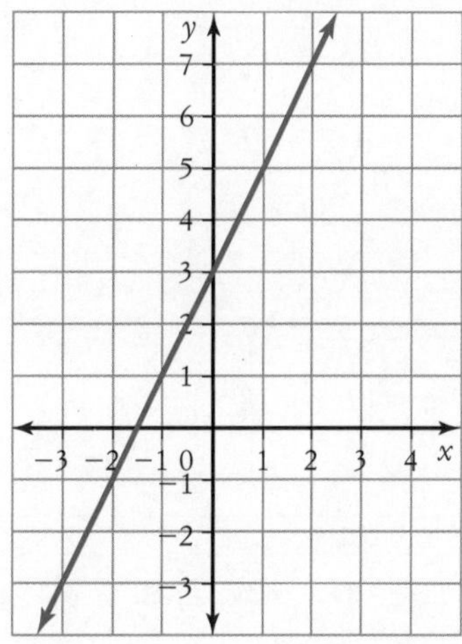

b Write the coordinates of the point where the line crosses the y-axis. ______

PART C: CHALLENGE

Bonus / 3 marks

Mr Black, Mr Brown and Mr Green were wearing a black tie, a brown tie and a green tie.

'Do you realise', said the man wearing the green tie, 'that not one of us is wearing the tie whose colour matches our name?'

'That's strange!' exclaimed Mr Black.

What colour tie was each man wearing?

C S F

12 FINDING THE RULE

FOR EACH TABLE OF VALUES, FIND THE RULE (FORMULA) AND COMPLETE THE TABLE.

a $y =$ ______

x	2	5	8	7	1	0
y	9	12	15	14		

b $y =$ ______

x	5	8	11	7	10	4
y	1	4	7	3		

c $y =$ ______

x	2	3	7	5	8	1
y	8	12	28	20		

d ______

h	12	6	2	10	4	8
t	6	3	1	5		

e ______

h	6	9	2	4	12	7
t	14	17	10	12		

f ______

h	5	3	11	9	4	8
t	2	0	8	6		

g ______

c	3	8	5	1	10	6
d	18	48	30	6		

h ______

c	12	10	7	8	16	13
d	7	5	2	3		

i ______

c	3	7	1	12	9	6
d	13	17	11	22		

j ______

k	2	3	4	5	6	7
p	6	9	12	15		

k ______

k	1	2	3	4	5	6
p	9	10	11	12		

l ______

k	1	2	3	4	5	6
p	4	6	8	10		

m ______

u	1	2	3	4	5	6
v	−1	2	5	8		

n ______

u	0	1	2	3	4	5
v	1	6	11	16		

o ______

u	4	5	6	7	8	9
v	4	6	8	10		

p ______

s	2	3	4	5	6	7
z	10	14	18	22		

q ______

s	1	2	3	4	5	6
z	6	9	12	15		

9780170454520

r ______

s	1	2	3	4	5	6
z	10	18	26	34		

s ______

g	3	4	5	6	7	8
n	7	8	9	10		

t ______

g	0	1	2	3	4	5
n	9	12	15	18		

u ______

g	1	2	3	4	5	6
n	7	14	21	28		

v ______

i	2	3	4	5	6	7
m	26	36	46	56		

w ______

i	1	2	3	4	5	6
m	1	7	13	19		

x ______

i	4	7	2	9	1	10
m	9	15	5	19		

y ______

i	2	7	11	8	4	5
m	1	16	28	19		

12 GRAPHING LINEAR EQUATIONS

GRAPH EACH LINEAR EQUATION AFTER COMPLETING THE TABLE OF VALUES.

WS WORKSHEET

1 $y = x + 3$

x	−1	0	1	2
y				

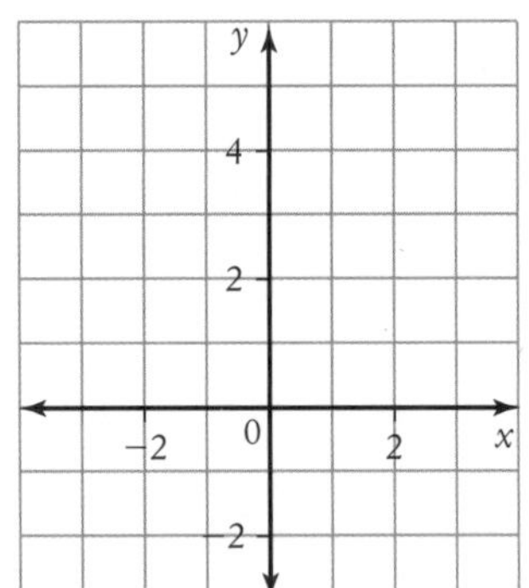

2 $y = 2x$

x	−1	0	1	2
y				

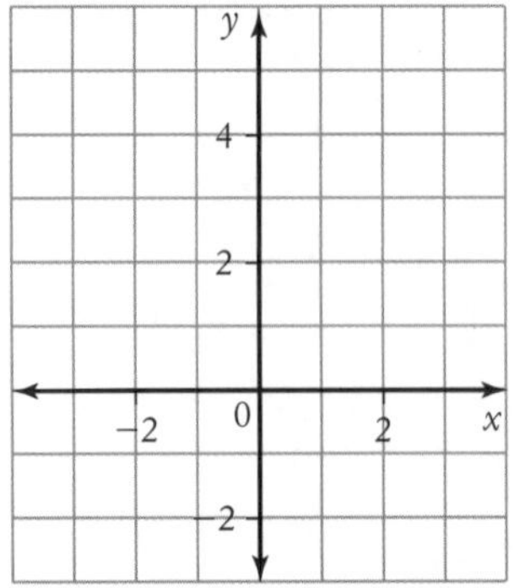

3 $y = \frac{1}{2}x + 1$

x	−1	0	1	2
y				

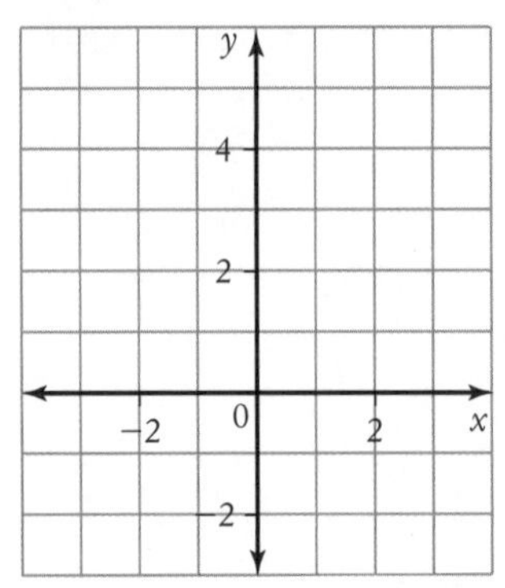

4 $y = 2x - 3$

x	−1	0	1	2
y				

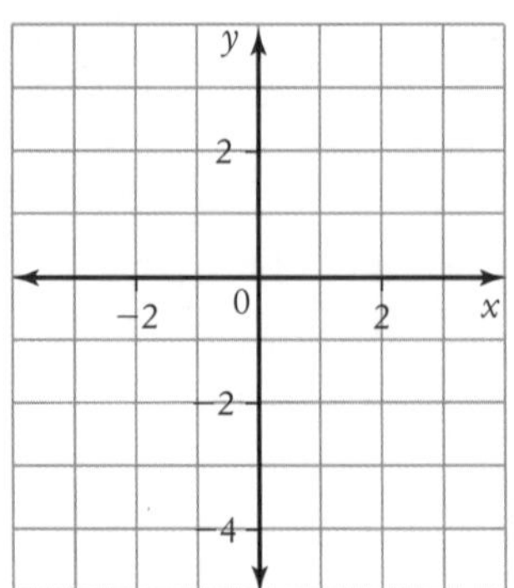

5 $y = 3x + 1$

x	−1	0	1	2
y				

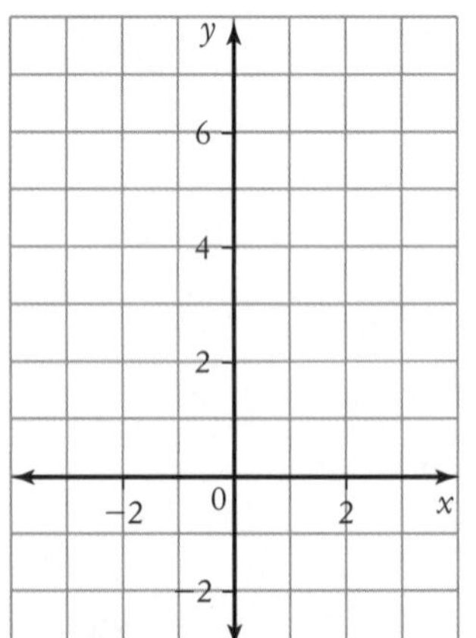

6 $y = -x + 5$

x	−1	0	1	2
y				

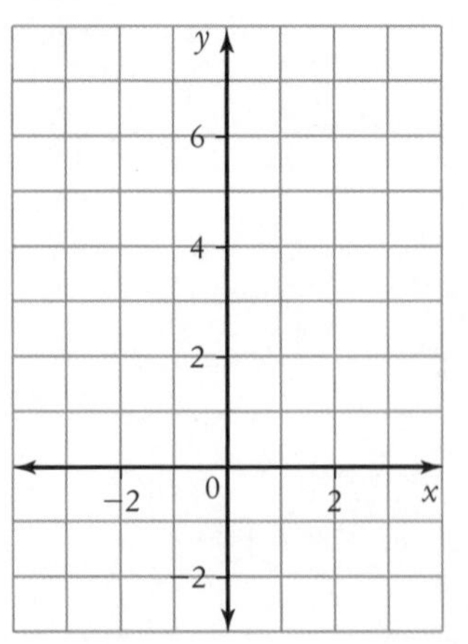

9780170454520

WS WORKSHEET

7 $y = -x - 1$

x	−1	0	1	2
y				

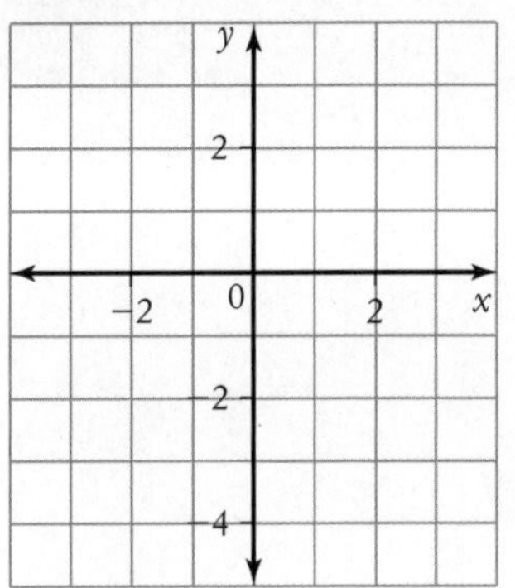

8 $y = -2x + 2$

x	−1	0	1	2
y				

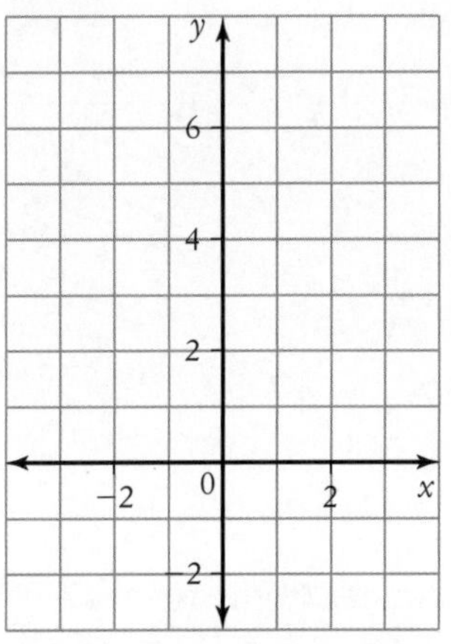

9 $y = -3x + 4$

x	−1	0	1	2
y				

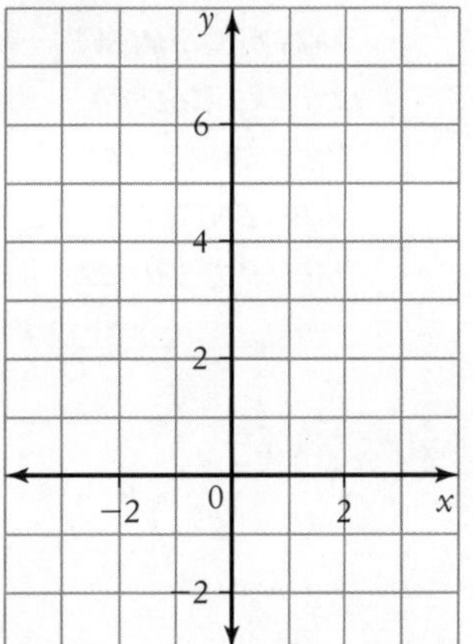

10 $y = x$

x	−1	0	1	2
y				

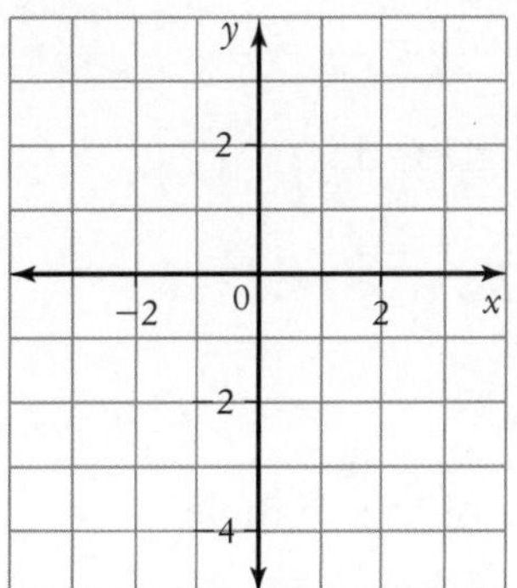

12 TABLES OF VALUES

THIS ASSIGNMENT COVERS ALGEBRA, THE NUMBER PLANE AND COMPLETING TABLES OF VALUES.

Name:

Due date:

Parent's signature:

Part A	/ 8 marks
Part B	/ 8 marks
Part C	/ 8 marks
Part D	/ 8 marks
Total	/ 32 marks

HW HOMEWORK

PART A: MENTAL MATHS

Calculators not allowed

1 List the factors of 20.

2 Decrease \$60 by 25%.

3 Factorise $9ab + 12b$. ______

4 **a** What type of triangle is this?

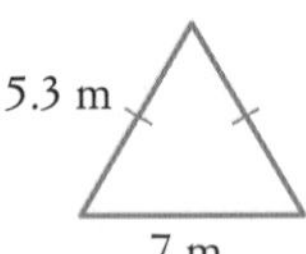

b Find the perimeter of this triangle.

c If the area of this triangle is 14 m^2, what is its perpendicular height?

5 Convert $\frac{5}{6}$ to a decimal.

6 Solve the equation $12 - 3m = 36$.

PART B: REVIEW

1 Complete each pattern.

a 7, 13, 19, 25, ______

b 11, 9, 7, 5, ______

2 If $u = -1$, evaluate $2u - 1$.

3 Complete:

a $3 \times 3 +$ ______ $= 14$

b $5 \times 2 -$ ______ $= 7$

4 On a number plane:

a what are the coordinates of the **origin**?

b what is another name for the horizontal axis?

c in which quadrant is the point $(7, -2)$ located?

C S F

9780170454520

PART C: PRACTICE

› The number plane
› Tables of values
› Finding the rule

1 (4 marks) Plot the points $A(-2, 1)$ and $B(2, -3)$ on the number plane and then write the coordinates of D and E.

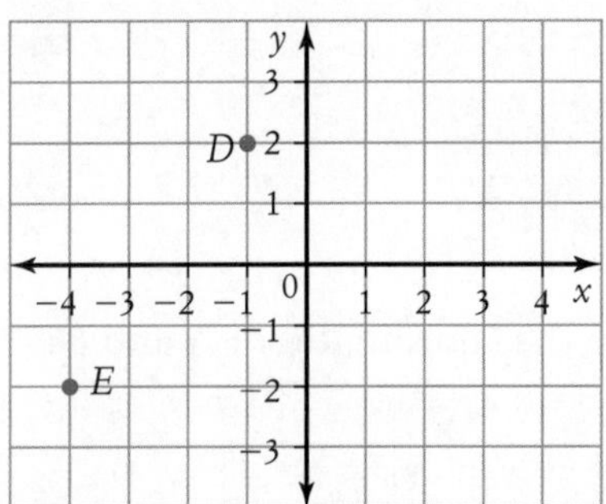

____

____

2 (2 marks) Complete this table of values using the formula $y = 3x + 4$.

x	−2	5	2	−1
y				

3 Find the rule connecting x and y for each table of values.

a

x	−1	0	1	2
y	3	4	5	6

____

b

x	−1	0	1	2
y	−9	−4	1	6

____

PART D: NUMERACY AND LITERACY

1 For the point $A(2, 3)$, what is the name given to the 3?

____

2 **a** What is a **quadrant** of the number plane?

____

____

b Show on the diagram how the quadrants are numbered.

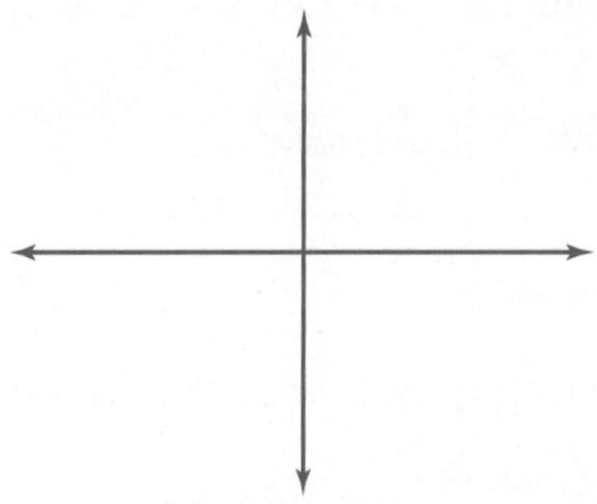

3 Which point on the number plane is 3 units right and 2 units down from (0, 0)?

____

4 Where on the number plane is the point (0, −4) located?

____

5 **a** Complete this table of values.

x	−1	0	1	2
y	3	7		15

b Find the rule connecting x and y in the above table.

____

6 Complete: All points in the 4th quadrant have a __________ x-coordinate and a __________ y-coordinate.

HW HOMEWORK

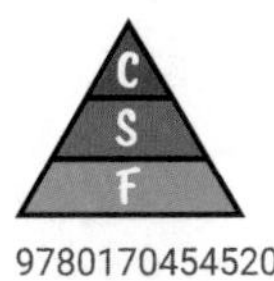

12 GRAPHING TABLES OF VALUES

Name:

Due date:

Parent's signature:

Part A	/ 8 marks
Part B	/ 8 marks
Part C	/ 8 marks
Part D	/ 8 marks
Total	/ 32 marks

THIS TOPIC SHOWS THAT NUMBER PATTERNS BECOME GEOMETRICAL PATTERNS WHEN GRAPHED ON A NUMBER PLANE. ALL THE POINTS SHOULD LINE UP.

PART A: MENTAL MATHS

Calculators not allowed

1 Evaluate $150 \div 3 - 12 \times 4$. ____________

2 Convert $\frac{1}{6}$ to a percentage.

____

3 Use a factor tree to write 90 as a product of its prime factors.

____

4 Solve the equation $48 - 6t = 12$.

____

____

5 Find the area of this kite.

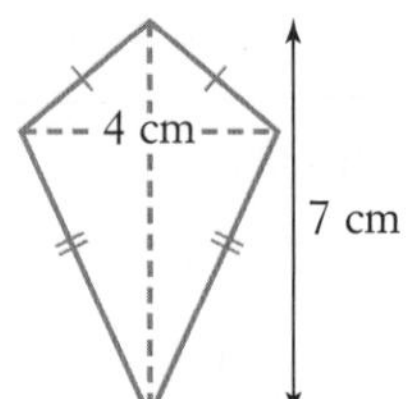

____

6 A movie starts at 3.45 p.m. and runs for 2 hours 13 minutes. What time does it finish?

____

7 Complete: $2350 \text{ mm}^2 =$ ________ cm^2

8 Find the value of x if these numbers have a mean of 6: 3, 7, 10, 6, 4, 6, x.

____

PART B: REVIEW

1 (2 marks) Write the coordinates of A and Q.

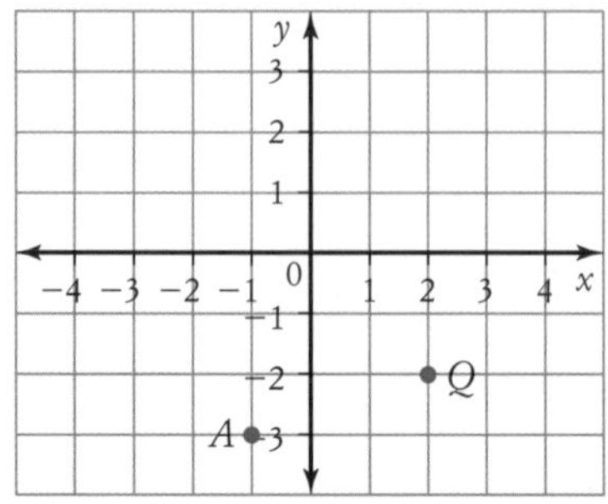

____

2 (2 marks) On the above number plane, plot the points $R(-4, 0)$ and $U(3, -1)$.

3 Write the coordinates of a point in the 2nd quadrant. ____________

4 Complete this table of values using the formula $y = 3x - 2$

x	−1	0	1	2
y				

5 Find the rule connecting x and y for each table of values.

a

x	0	1	2	3
y	0	4	8	12

____

b

x	−1	0	1	2
y	−4	−2	0	2

____

C S F

HW HOMEWORK

9780170454520

PART C: PRACTICE

› Finding rules for number patterns
› Graphing tables of values

1 **a** Complete the table for the number pattern.

n	1	2	3	4	5	6
Term, T	7	10	13	16		

b Find a formula for the nth term, T_n

c Use the formula to find the 20th term.

2 **a** Complete the table for the toothpick pattern.

 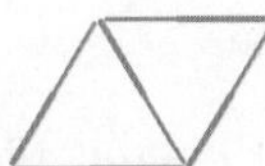 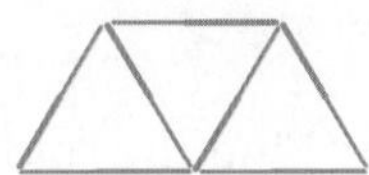

No of triangles, x	1	2	3	4	5
No of toothpicks, y	3	5			

b Find a formula for y.

c How many toothpicks are needed to make 15 triangles?

d Graph the table of values from part **a** on this number plane.

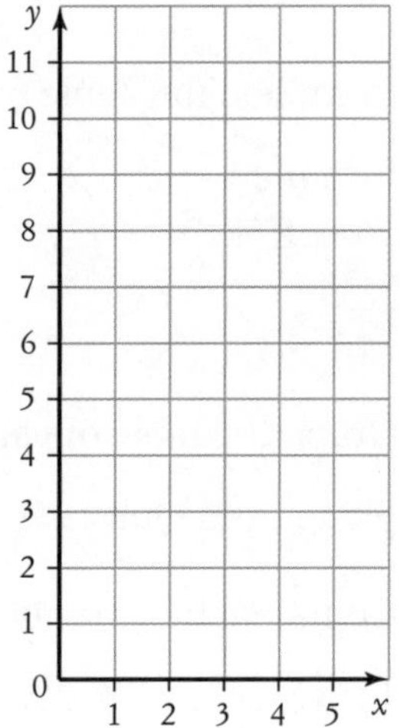

3 Also graph this table of values on the above number plane.

x	0	1	2	3
y	7	5	3	1

C S F

PART D: NUMERACY AND LITERACY

1 If a point on the number plane has:

a negative x- and y-coordinates, what quadrant is it in?

b a y-coordinate of 0, what axis is it on?

2 **a** Complete the table for the toothpick pattern.

 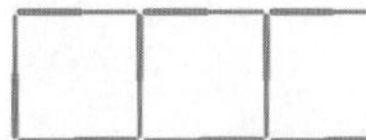

No of squares, x	1	2	3	4	5
No of toothpicks, y					

b What is the pattern in the y-values?

c Explain how to find the formula for y in terms of x.

d How many toothpicks are needed to make 12 squares?

3 What is the name of the point with x- and y-coordinates of 0?

4 Find the rule for this table of values.

x	0	1	2	3
y	0	1	2	3

HW HOMEWORK

12 GRAPHING LINEAR EQUATIONS

BY NOW, YOU SHOULD BE MASTERING GRAPHING EQUATIONS ON THE NUMBER PLANE, AFTER COMPLETING A TABLE OF VALUES.

Name:

Due date:

Parent's signature:

Part A	/ 8 marks
Part B	/ 8 marks
Part C	/ 8 marks
Part D	/ 8 marks
Total	/ 32 marks

PART A: MENTAL MATHS

Calculators not allowed

1 Evaluate each power:

a $(-2)^4$ ____________

b 5^0 ____________

2 Find the highest common factor of 12 and 16.

____

3 Simplify $4a - 3 + 6 - a$ ____________

4 Find the area of this trapezium.

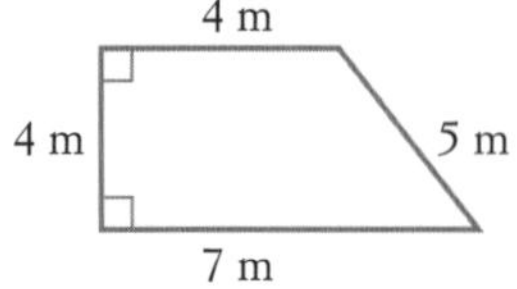

5 If a die is rolled 48 times, how many times should you expect a '1' to come up?

____

6 Find the value of x if these numbers have a range of 8 :

3, 7, 10, 6, 4, 6, x.

____

7 Find m.

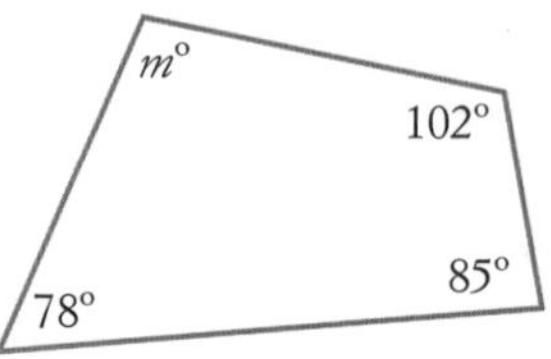

____

PART B: REVIEW

1 **a** Complete this table of values.

x	−1	0	1	2
y	−4	−1	2	

b Find the rule for the table of values.

____

c Graph the table of values on the number plane.

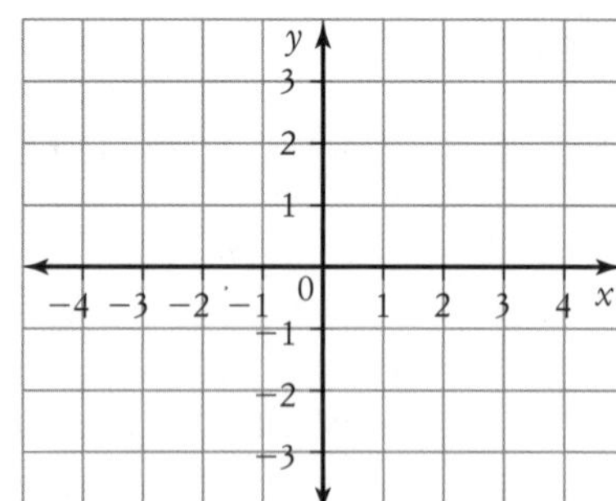

2 **a** Complete this table of values using the formula $y = 2 - x$.

x	−1	0	1	2
y				

b Graph the table of values on the number plane from Question **1c** above.

3 **a** Complete this number pattern:

14, 19, 24, 29, 34, 39, ________

b Find a formula for the nth term, T.

____

c Use the formula to find the 100th term.

____

C S F

9780170454520

PART C: PRACTICE

› Graphing linear equations
› Vertical and horizontal lines
› Solving linear equations graphically

1 (6 marks) Graph each linear equation on the number plane.

a $y = 2x + 1$

b $y = x - 3$

c $x = -1$

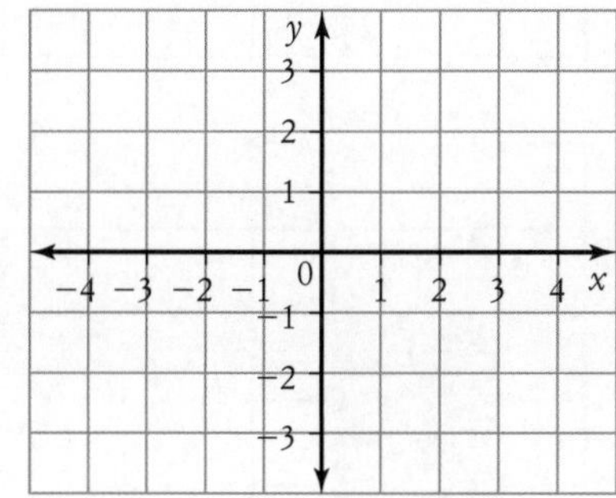

2 (2 marks) Solve the equation $3x - 3 = 3$ graphically.

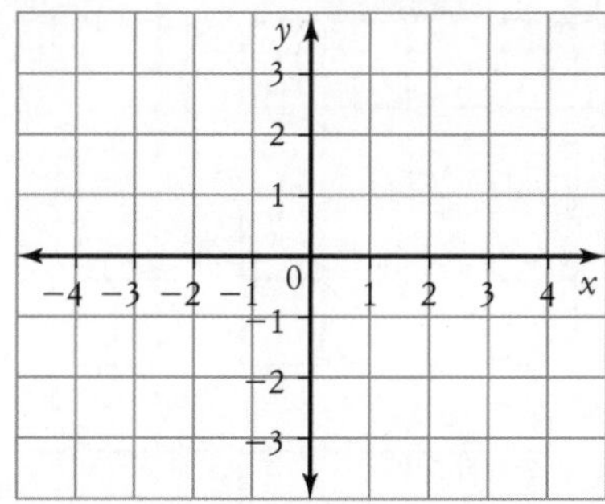

PART D: NUMERACY AND LITERACY

1 (5 marks) Graph $y = 1 - x$ and $y = 2x - 2$ on the number plane and write the coordinates of the point where the 2 lines intersect.

2 Find the equation of:

a the horizontal line that passes through (0, 2)

____

b the vertical line 3 units left of the y-axis

____

3 Explain why the point (3, 5) lies on the graph of $y = 3x - 4$.

____

____

____

____

12 LINEAR EQUATIONS CROSSWORD

The words for completing this crossword puzzle are listed below in alphabetical order. Arrange them in the correct places in the puzzle. (*Note*: Ignore the hyphens in the words.)

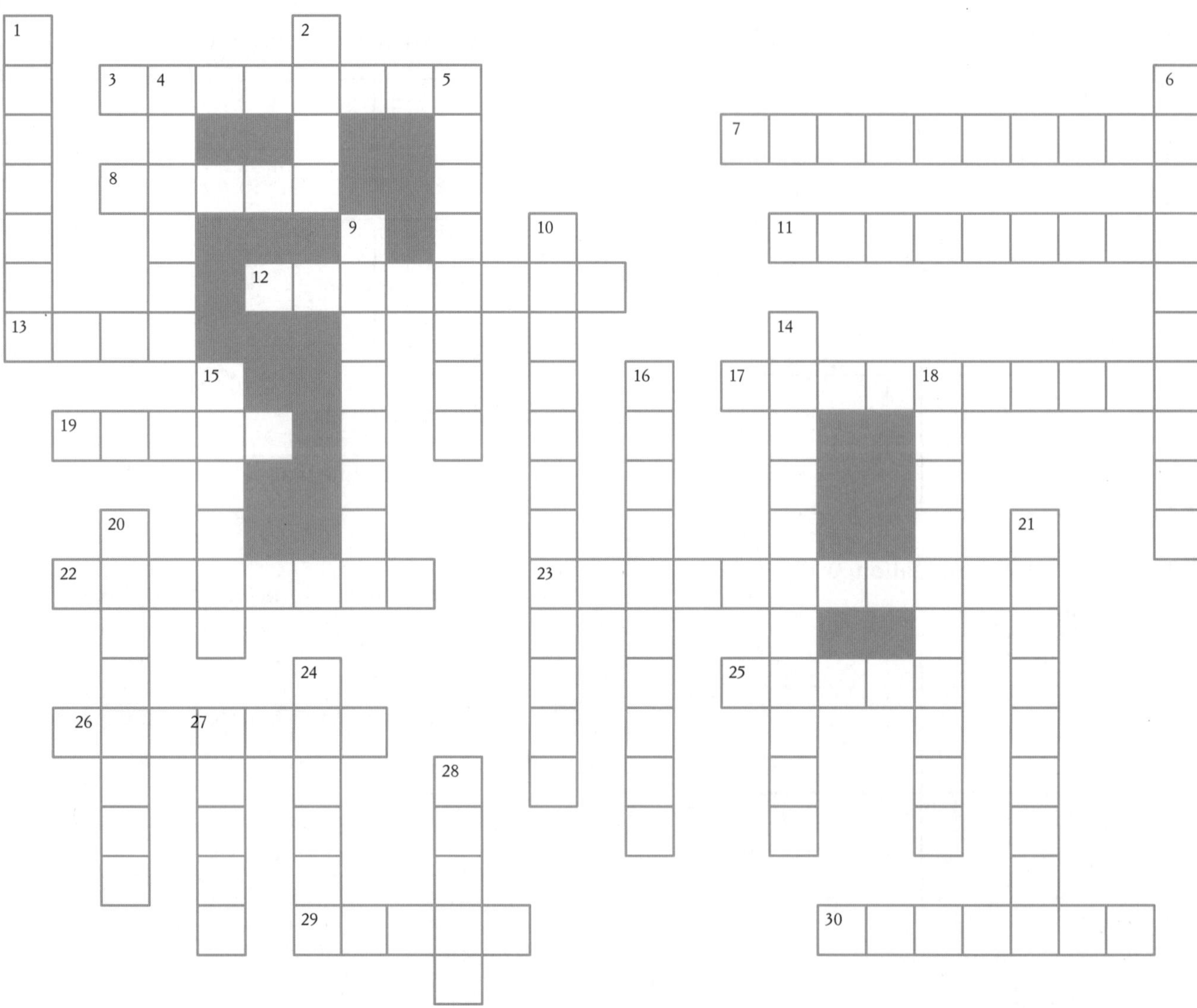

ARROWS	AXES	COEFFICIENT
CONSTANT	CONTINUOUS	COORDINATES
DECREASING	EQUATION	EVALUATE
FORMULA	GRAPH	INCREASING
INFINITE	INTERSECTION	LINEAR
NON-LINEAR	PARALLEL	PATTERN
PLANE	RULE	SATISFY
SLOPE	STEEPNESS	SUBSTITUTE
TABLE	VALUES	VARIABLE
X-AXIS	Y-AXIS	Y-INTERCEPT

9780170454520

Chapter 1

StartUp assignment 1 PAGE 01

1 72
2 **a** 48 cm **b** 96 cm^2
3 {head, tail}
4 **a** $4m$ **b** $8h$
5 135°
6 **a** $x = -9$ **b** $d = 7$
7 42.5% **8** $\frac{1}{3}$
9 −5 **10** 1875 g
11 9 : 4 **12** 428
13 **a** 40 **b** 289
c 729 **d** 10.24
e 256 **f** 21
g 26 **h** 6
14

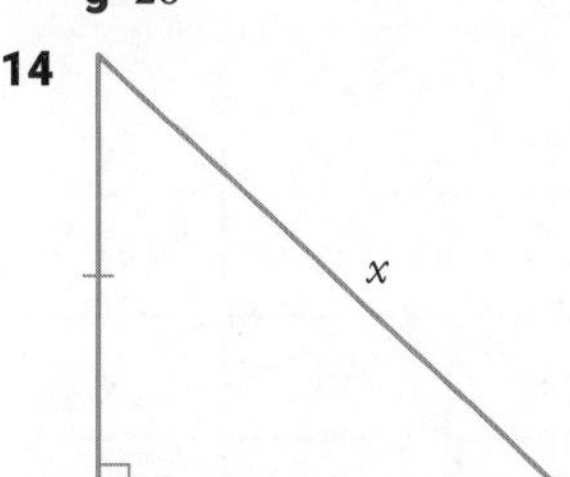

15 **a** True **b** False
c False **d** True
e False **f** True
16 49, 225, 81
17 **a** 4.90 **b** 10.39
c 6.32 **d** 5.42
18

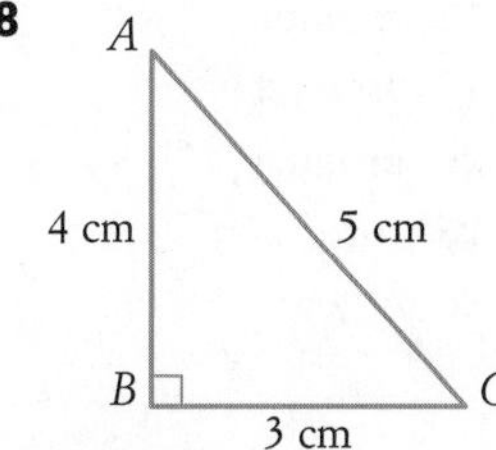

Challenge

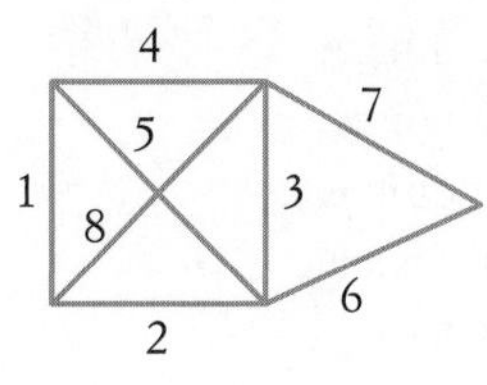

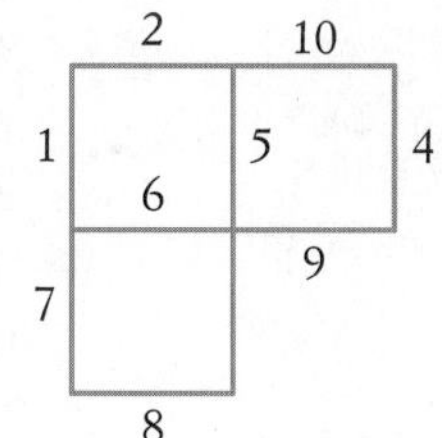

Pythagoras' discovery PAGE 03

1 $26^2 = 24^2 + 10^2$ **2** $65^2 = 60^2 + 25^2$
3 $29^2 = 20^2 + 21^2$ **4** $30^2 = 24^2 + 18^2$
5 $35^2 = 28^2 + 21^2$ **6** $34^2 = 30^2 + 16^2$
7 $15^2 = 12^2 + 9^2$ **8** $85^2 = 75^2 + 40^2$
9 $75^2 = 72^2 + 21^2$ **10** $50^2 = 48^2 + 14^2$
11 $5^2 = 4^2 + 3^2$ **12** $25^2 = 24^2 + 7^2$
13 $97^2 = 72^2 + 65^2$ **14** $13^2 = 12^2 + 5^2$
15 $6^2 = 4.8^2 + 3.6^2$ **16** $2^2 = 1.6^2 + 1.2^2$
17 $2.5^2 = 2^2 + 1.5^2$ **18** $4.1^2 = 4^2 + 0.9^2$
19 $51^2 = 45^2 + 24^2$ **20** $10^2 = 8^2 + 6^2$
21 $17^2 = 15^2 + 8^2$ **22** $100^2 = 96^2 + 28^2$

Pythagorean triads PAGE 04

1 **a** 7.5 cm **b** 13 cm
4 **a** (7, 24, 25) **b** (8, 6, 10)
c (20, 48, 52)
5 **a** (7, 24, 25) **b** (11, 60, 61)
c b and c differ by 1
6 **a** (9, 40, 41) **c** $a^2 = b + c$
d $c - b = 1$

Challenge 12 cubits above ground level

Applications of Pythagoras' theorem PAGE 06

1 17.89 m
2 **a** No, 60 cm **b** No, 84.85 cm
c No, 84.85 cm **d** Yes, 103.92 cm
3 102.08 cm, 128.02 cm, 152.69 cm
4 5.66 m **5** 120 nautical miles
6 **a** 37.17 cm **b** 43.27 cm
7 21.36 cm **8** 1.33 m
9 **a** 6.5 cm
b shorter sides 2 cm, 4 cm (other answers possible)
10 **a** 363.85 m **b** 834.07 m
11 19.43 m
12 $h = 8$ cm, $d = 8.25$ cm **13** $JL = \sqrt{8} = 2.83$ cm

Chapter 2

StartUp assignment 2 PAGE 11

1 9 : 05 p.m. **2** 1040
3 vertically opposite **4** 3rd
5 5.309 **6** 30 cm
7 1000 **8** $7x$
9 hexagon **10** 25, 36
11 4 **12** 6 cm^2
13 5 **14** 1, 2, 4, 5, 10, 20
15 **16** 0.6
17 $^-9$, $^-8$, $^-5$, 0, 4, 17 **18** 4
19 **a** 145.24 **b** 145
20 **a** 64 **b** 9
c −1 **d** −11
e 320 **f** 8
21 47 **22** 18

23 3.416, 3.41, 3.146, 3.14

24 $\frac{3}{20}$

25 **a** 4 **b** −10
c 96 **d** 60
e 81 **f** 14

26 253

27 24

28

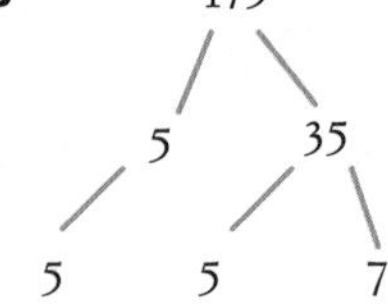

$175 = 5^2 \times 7$

Challenge

4	9	2
3	5	7
8	1	6

Decimal number grids

PAGE 13

1

+	**6.8**	**3.1**
5.6	12.4	8.7
4.4	11.2	7.5

2

+	**8.3**	**2.6**
10.7	19	13.3
2.9	11.2	5.5

3

+	**4.4**	**5.6**
1.2	5.6	6.8
2.6	7	8.2

4

+	**1.12**	**3.46**
2.05	3.17	5.51
7.36	8.48	10.82

5

+	**8.7**	**1.9**	**4.8**
2.2	10.9	4.1	7
5.8	14.5	7.7	10.6

6

+	**1.15**	**2.64**	**2.06**
1.07	2.22	3.71	3.13
7.96	9.11	10.60	10.02

7

−	**8.4**	**7.7**
2.5	5.9	5.2
3.8	4.6	3.9

8

−	**9.7**	**6.3**
1.4	8.3	4.9
4.9	4.8	1.4

9

×	**1.2**	**2.4**	**3.6**	**3.5**
0.2	0.24	0.48	0.72	0.7
0.3	0.36	0.72	1.08	1.05
0.7	0.84	1.68	2.52	2.45

10

÷	**8.4**	**9.6**	**7.2**	**3.6**
0.2	42	48	36	18
0.4	21	24	18	9
1.2	7	8	6	3
0.6	14	16	12	6

Numbers crossword

PAGE 14

Across

1 grouping
8 ascending
11 factor
12 multiple
14 decimal
17 greatest
18 one
20 product
22 tree
23 surd
28 notation
30 order
31 places

Down

2 right
3 index
4 terminating
5 integer
6 square root
7 points
8 add
9 difference
10 common
13 evaluate
15 lowest
16 left
19 zero
21 recurring
24 divisor
25 subtract
26 prime
27 mental
29 power

9780170454520

Chapter 3

StartUp assignment 3 PAGE 30

1 4200 **2** 23 or 29

3 any 6-sided shape such as 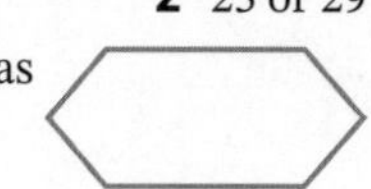

4 110° **5** 0.882

6 a triangular pyramid **b** 4

7 Origin **8** 7

9 36 **10** 5

11 75% **12** False

13 a 26 m **b** 30 m^2

14 a −28 **b** 8 **c** 7 **d** 36 **e** 10 **f** 10

15 1, 2, 4, 8, 16 **16** 54

17 240 **18** 6

19 a 33 **b** −1

20 a $2h$ **b** r^4 **c** $6x - 4$ **d** $24x$

21 a $5y$ **b** $n + 1$

22 18

23 a true **b** true

24 38 **25** 36

26 7

27 Teacher to check, 60 $2^2 \times 3 \times 5$

Challenge 61

What's the expression? PAGE 32

1 Z **2** Q
3 E **4** Y
5 C **6** S
7 V **8** W
9 J **10** O
11 T **12** U
13 P **14** N
15 T **16** I
17 E **18** P
19 X **20** H
21 M **22** B
23 F **24** N
25 D **26** K
27 L **28** A
29 G **30** R

I'll give you two expressions: one is 'I'd be happy as Larry'. The other is 'I've got Buckley's chance'!

Collecting like terms PAGE 34

1 $6x$ **2** $11y$
3 $5a$ **4** $5t$
5 $9ab$ **6** $20d$
7 $10st$ **8** $3x$
9 $9a^2$ **10** t
11 $8ab$ **12** $2ty$
13 mp **14** $2x$
15 $7w$ **16** $-3c$
17 0 **18** $4m$
19 $-e$ **20** $2h + 6$
21 $3y$ **22** $8c - 2d$
23 $10p + 9$ **24** $6s$
25 $5y^2$ **26** $3d$
27 $2a + 2f$ **28** $3c + 7d$
29 $7h + 7$ **30** $5u^2$
31 $6a - 10b$ **32** $7g - 1$
33 $3x + b$ **34** $20m - 5n$
35 $11a^2 + a$ **36** $2y$
37 $6w + 7wt$ **38** $7a + 2b$
39 $2x^2 + 4y$ **40** $8f + 4$
41 $3y^2 + 7x$ **42** $k + 2p$
43 y **44** $2y - 1$
45 $y + 2m$ **46** $3q + 1$
47 $8r^2 - 3r$ **48** $6a - b$
49 $8t + v$ **50** $3k + 8$
51 $17b - 8$ **52** $y + 3z$
53 $-x + 3$ **54** $3ab + 3b$
55 $p^2 + 4p$ **56** $2u^2 + 1$
57 $-2a + 5$ **58** $7x - 8$
59 $9x + 3$ **60** $2p - 5q$
61 $6w$ **62** $18x$
63 $2p - 3a$ **64** $c + 6d$
65 $7jk - 3k$ **66** $-7b$
67 $3x$ **68** $-7y^2 + 4yz$
69 $-4y + 7$ **70** $3w^2 + 3u$
71 $bc + 7cd$ **72** $7f^2 + 2ef$

Factorising puzzle PAGE 35

Al-Khwarizmi

Chapter 4

StartUp assignment 4 PAGE 45

1 2 hours and 12 minutes **2** −19

3 $\frac{5}{3} = 1\frac{2}{3}$ **4** Prime

5 $-18a + 12$ **6** 21

7 **b** = 5

8 a 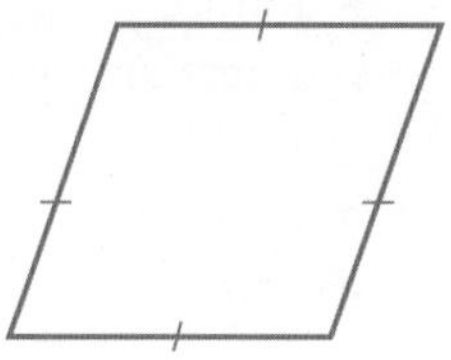 **b** 2

9 $y = 17$ **10** −5, −3, 0, 1, 5

11 7.025 **12** $-24b^2d$

13 Add its digits and see if the sum is a multiple of 9

14 $\frac{3}{10}$ **15** 180°

16

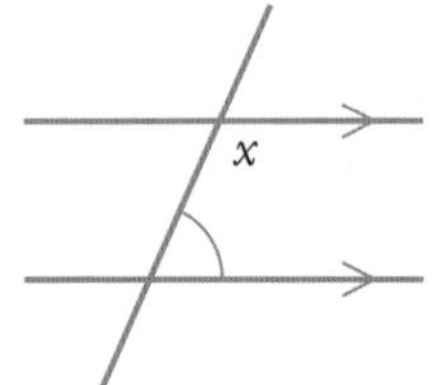

17

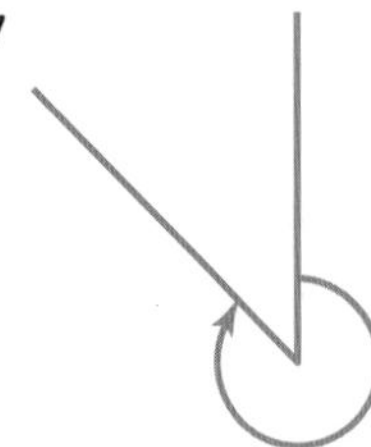

18 Supplementary

19 $r = 130, s = 130$

20

21 Trapezium

22 **a** Vertically opposite angles **b** $p = 105, q = 75$

23 360°

24 **a** Equilateral triangle **b** 60°

25

26 rhombus, kite or square

27 **a** Parallelogram **b** 0
c 2

28 360°

29

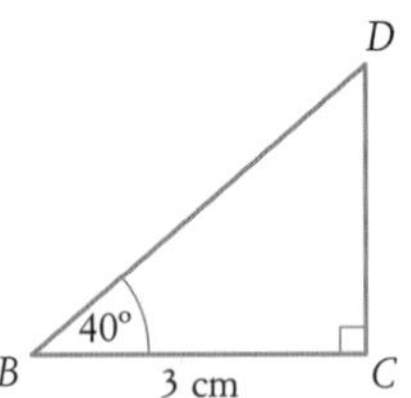

30 **a** *PS* **b** *SQ*
c $\angle PQR$

Challenge

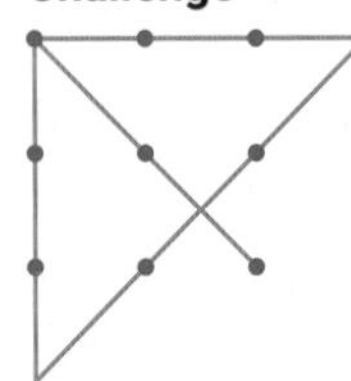

Classifying quadrilaterals

PAGE 50

1 Parallelogram **2** Rhombus
3 Rectangle **4** Kite
5 Square **6** Trapezium
7 Parallelogram
8 **a** True **b** True
c False **d** True
e False
9 **a** True **b** True
c True **d** True
e False
10 **a** False **b** True
c False **d** False
11 **a** False **b** False
c False **d** True
12 **a** Rhombus **b** Parallelogram
c Parallelogram **d** Parallelogram
e Parallelogram **f** Rhombus
g Rectangle **h** Rectangle
i Parallelogram **j** Kite

Challenge

a False **b** True

Find the unknown angle

PAGE 52

1 $a = 142, b = 38$ **2** $x = 48$
3 $y = 202$ **4** $a = 100$
5 $c = 115$ **6** $x = 72$
7 $a = 70$ **8** $p = 20$
9 $a = 50, b = 140$ **10** $x = 22$
11 $a = c = 114, b = d = 66$ **12** $h = 50, k = 75$
13 $x = 45$ **14** $k = 40$
15 $r = 30, s = 60, t = 150$ **16** $m = 55, n = 55$
17 $x = 95, y = 85, z = 95$ **18** $z = 65, x = 65, y = 50$
19 $p = 120$ **20** $x = 130$

Geometry crossword

PAGE 54

Across

4 rectangle **5** rhombus
6 bisect **9** corresponding
11 exterior **13** vertex
14 isosceles **16** acute
17 perpendicular **20** co-interior
21 obtuse **22** alternate
23 parallel **24** right
25 supplementary **26** angle
27 vertically

Down

1 trapezium **2** set
3 ruler **7** construct
8 kite **9** complementary
10 parallelogram **12** quadrilateral
15 compasses **17** protractor
18 revolution **19** equilateral
25 sum

Chapter 5

StartUp assignment 5

PAGE 62

1 **a** 432 **b** 5
c $\frac{1}{9}$ **d** 32
2 $u = 130$ **3** $p - 2$
4 $\frac{1}{4}$ **5** Both 25°
6 $5a - 3$ **7** $0.\dot{3}$
8 3^4
9 **a** 54 cm^2 **b** 27 cm^3
10 $4y(3x - y)$ **11** False
12 **a**

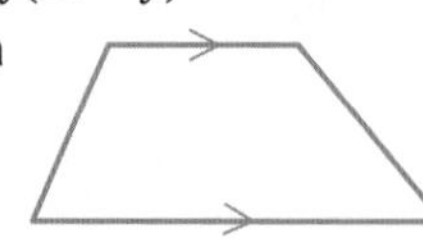

b

13 **a** 12 m **b** 9 m^2

14 Triangular prism

15 **a** 1800 **b** 3.25

c 960 **d** 2.4

e 10 000 **f** 1

16 360°

17 **a** 6.32 cm **b** 6 cm^2

18 **a** **b** 2

19 **a** 45.24 **b** 162.86

20 **a** 26 cm **b** 27 cm^2

21 96 cm^3 **22** 18 cm^2

23 2.2 cm

24 **a** $2w + 2l$ **b** lw

Challenge

Composite areas PAGE 64

1 30 cm^2	**2** 36 cm^2
3 24 cm^2	**4** 78 cm^2
5 30 cm^2	**6** 13 cm^2
7 40 cm^2	**8** 36 cm^2
9 27.5 cm^2	**10** 28 cm^2
11 84 cm^2	**12** 50 cm^2
13 29.5 cm^2	**14** 1600 cm^2
15 34 m^2	**16** 2575 mm^2
17 19.85 m^2	**18** 3 070 000 mm^2

Parts of a circle PAGE 66

2 **a** Diameter **b** Segment

c Arc **d** Quadrant

e Chord **f** Tangent

g Radius **h** Sector

i Semi-circle **j** Segment

k Sector **l** Circumference

A page of circles PAGE 67

(circumference, area)

1 12.57 cm, 12.57 cm^2	**2** 9.42 cm, 7.07 cm^2
3 50.27 cm, 201.06 cm^2	**4** 6.28 cm, 3.14 cm^2
5 15.71 cm, 19.63 cm^2	**6** 43.98 cm, 153.94 cm^2
7 3.14 cm, 0.79 cm^2	**8** 31.42 cm, 78.54 cm^2
9 18.85 cm, 28.27 cm^2	**10** 35.81 cm, 102.07 cm^2
11 62.83 cm, 314.16 cm^2	**12** 37.70 cm, 113.10 cm^2
13 56.55 cm, 254.47 cm^2	**14** 53.41 cm, 226.98 cm^2
15 21.99 cm, 38.48 cm^2	**16** 34.56 cm, 95.03 cm^2
17 23.88 cm, 45.36 cm^2	**18** 28.90 cm, 66.48 cm^2

Chapter 6

StartUp assignment 6 PAGE 77

1 145° **2** 9

3 Scalene **4** 5 h 50 min

5 $2x + 5$ **6** $\frac{4}{15}$

7 3rd **8** 12 cm^2

9 $-14a - 7$ **10** 0 or 5

11 11 **12** -10

13 $5y - 2x$ **14** 4

15 No

16 0.3, 0.31, 0.317, 0.37, 0.371

17 1, 2, 4, 5, 10, 20, 25, 50, 100

18 **a** $\frac{9}{50}$ **b** $\frac{18}{25}$

c $\frac{7}{20}$

19 **a** 0.085 **b** 90

c 60 **d** $27.30

20 $\frac{1}{2}$ **21** $45

22 45 **23** 75%

24 **a** 0.54 **b** 0.375

c 0.583

25 $480

26 **a** 120 **b** 5400

c 270

27 **a** $\frac{17}{20}$ **b** $\frac{6}{25}$

c $\frac{2}{5}$

28 46 **29** 54%

Challenge CC, C, A, C, CC, C, A, C, CC

Percentage shortcuts PAGE 80

1 **a** 0.12 **b** 0.73

c 0.05 **d** 0.4

e 0.186 **f** 0.08

g 0.031 **h** 1.22

i 0.0695 **j** 0.125

k 1.5 **l** 0.0825

2 **a** $14.40 **b** $397.71

c $2.52 **d** $15.46

e $142.40 **f** $11 926

g 74c **h** $4.53

i $53.28 **j** $130

k $63.72 **l** $3.83

3 **a** $90.85 **b** $37.50

c $179.20 **d** $360.40

e $24.75 **f** $414.75

g $82.94 **h** $255.20

i $408.62 **j** $47.60

k $572.47 **l** $91

4 **a** $27 **b** $69

c \$760 **d** \$152.15
e \$77.88 **f** \$377.20
g \$5208 **h** \$577.50
i \$72.11 **j** \$222.78
k \$22.78 **l** \$254.67

5 \$437.50

6 a \$65.65 **b** \$11.26
c \$33

7 \$84 944.65

8 a 1.07 **b** \$5350
c \$5724.50 **d** \$6125.22
e The size of the investment after *n* years

Percentages without calculators PAGE 82

1 a $\frac{4}{5}$ **b** $\frac{1}{4}$
c $\frac{3}{200}$ **d** $\frac{13}{25}$

2 a 0.14 **b** 0.125
c 0.66 **d** 0.6

3 a 8 **b** 14
c 18 **d** 25
e \$77 **f** 42 min
g 750 g **h** 800 mL

4 a $\frac{1}{50}$ **b** $\frac{2}{3}$
c $\frac{11}{25}$ **d** $\frac{17}{20}$

5 a 0.78 **b** 0.08
c 0.5625 **d** 0.3209

6 a 0.2875 **b** 0.162
c 3.404 **d** 57
e 2400 **f** 500

7 a 70% **b** 20%
c $33\frac{1}{3}$% **d** 75%
e 29% **f** 90%
g 9.1% **h** 47.5%

8 a 5 **b** 28
c 12 seconds **d** 6 months
e 7 **f** 9
g \$6 **h** \$5.60
i 8 hours **j** 45 minutes
k \$3.54 **l** 1.2 metres

9 a $\frac{1}{6}$ **b** $\frac{7}{10}$
c $\frac{1}{5}$ **d** $\frac{3}{25}$

10 87%, 0.8, 0.78, $\frac{3}{4}$ **11** 27

12 140

13 a \$80 **b** \$45
c \$30.80

14 a 60% **b** 33%
c 75% **d** 12.5%

15 a 0.12 **b** 6.4
c 1.4 **d** 0.3

16 a \$30 **b** \$294
c \$119

17 500 **18** 80%

19 \$20 000 **20** \$69

21 80 **22** $12\frac{1}{2}$%

23 20%

Fractions, decimals and percentages PAGE 92

1 a 0.17 **b** 0.25
c 0.09 **d** 0.38
e 0.875 **f** 0.7
g 0.041 **h** 0.637
i 0.005 **j** 0.101

2 a 15% **b** 42.7%
c 8% **d** 80%
e 1.5% **f** 72.4%
g 69% **h** 30.1%
i 0.7% **j** 28.5%

3 a $\frac{19}{25}$ **b** $\frac{33}{50}$
c $\frac{1}{20}$ **d** $\frac{33}{100}$
e $\frac{2}{5}$ **f** $\frac{24}{25}$
g $\frac{5}{8}$ **h** $\frac{49}{100}$
i $\frac{1}{6}$ **j** $\frac{17}{200}$

4 a 75% **b** 65%
c 60% **d** $87\frac{1}{2}$%
e 76% **f** $17\frac{1}{2}$%
g $66\frac{2}{3}$% **h** $63\frac{1}{3}$%
i $91\frac{2}{3}$% **j** $68\frac{3}{4}$%

5 a $\frac{3}{5}$ **b** $\frac{1}{50}$
c $\frac{9}{50}$ **d** $\frac{17}{20}$
e $\frac{7}{100}$ **f** $\frac{283}{500}$
g $\frac{18}{25}$ **h** $\frac{1}{8}$
i $\frac{6}{125}$ **j** $\frac{2}{5}$

6 a 0.625 **b** 0.45
c 0.86 **d** 0.3
e 0.16 **f** 0.8
g 0.34 **h** 0.8375
i 0.857142 **j** 0.28

7 Fractions: $\frac{1}{10}, \frac{1}{3}, \frac{1}{5}, \frac{1}{2}, \frac{1}{4}, \frac{2}{3}, \frac{3}{5}$
Decimals: 0.1, 0.75, 0.2, 0.05, 0.25, 0.6, 0.125
Percentages: $33\frac{1}{3}$%, 75%, 5%, 50%, $12\frac{1}{2}$%, 60%

Chapter 7

StartUp assignment 7

PAGE 93

1 $66\frac{2}{3}$ (or $66.\dot{6}\%$)
2 12.85 cm
3 1000
4 \$175
5 $8x - 4y$
6 360°
7 $\frac{2}{7}$
8 30 m^2
9 −3
10 Yes
11 $1\frac{1}{4}$
12 16.12 cm
13 2^{10}
14 20%
15 36 cm^3
16 a 6, 7, 10, 10, 11, 13, 13, 13, 16
b 11
c 13
d 10
e 11
17 a 20
b 2
c 5
d 5
e 15%
18 a 18.5
b 23
19 a 24
b $\frac{1}{6}$
c 70.8%
d 4
e 6
f 1
20 Teacher to check, for example, 7, 8, 10, 12, 12.
21 a 24
b 74 km/h
c 2
d 24
e $\frac{3}{8}$

Challenge Any two amounts that add to \$13 400.

Data match-up

PAGE 95

Categorical data: level of First Aid training, make of car, method of travel to work, film classification rating, type of home lived in, brand of toothpaste, outcome of a coin toss, attitude towards a new national flag, colour of eyes, marital status, month of birth, favourite radio station, setting of an air conditioner, exam grade, day of week a person shops, favourite sport

Numerical: discrete: dress size, traffic crossing bridge in an hour, goals scored by a hockey team, crowd size at a football match, exam mark as a whole percentage, number of times a person streamed a film last week, sum of 2 dice rolled, word length of an essay, year of birth (could also be categorical), school population, price of an ice cream, number of phones owned, house number (could also be categorical), number of apps on phone, number of bedrooms in home

Numerical: continuous: volume of air in a balloon, intensity of a light, running time of an athlete, height of a tree, available memory on a computer, amount of chlorine in a pool, loudness of a lawn mower, floor area of an office, mass of a truck, speed of a car, temperature of an oven, amount of electricity used daily, distance lived from school, a person's blood alcohol level, pulse rate in beats/minute, capacity of a car's fuel tank, length of a phone call

Mean, median, mode

PAGE 96

Enthusiasm is the greatest asset in the world. It beats money, power and influence.

Data crossword

PAGE 98

Across

3 six
4 categorical
5 survey
6 five
7 eight
8 bias
9 frequency
12 plot
14 sample
16 mean
17 histogram
21 polygon
22 line
24 median
27 outlier
28 spread
29 two
31 dot
32 number

Down

1 range
2 random
3 statistics
4 cluster
10 numerical
11 stem
13 location
15 often
18 analyse
19 census
20 population
23 average
24 mode
25 data
26 order
30 odd

Chapter 8

StartUp assignment 8

PAGE 106

1 $\frac{1}{8}$, 17%, $\frac{7}{40}$, 0.2
2 42 cm^3
3 $x = 24$
4 −27
5 $7d - 16$
6

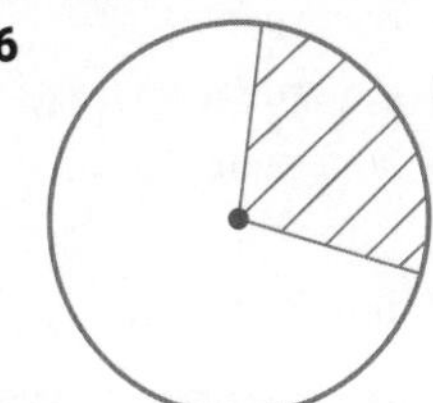

7 a Dot plot
b 6
c 4
8 1 h 3 min
9 $C = 2\pi r$
10 $-12m^2n$
11 $0.\dot{2}$
12 60 cats
13 301.59 cm^3
14 rectangle
15 Teacher to check
16

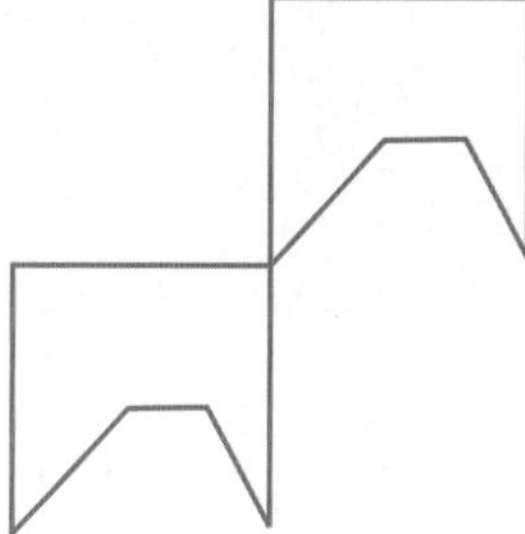

17 Yes
18 65°
19 a Quadrilateral with 4 equal sides
b 2
c *AB* and *DC* (or *AD* and *BC*)
d *AC* and *DB*

20

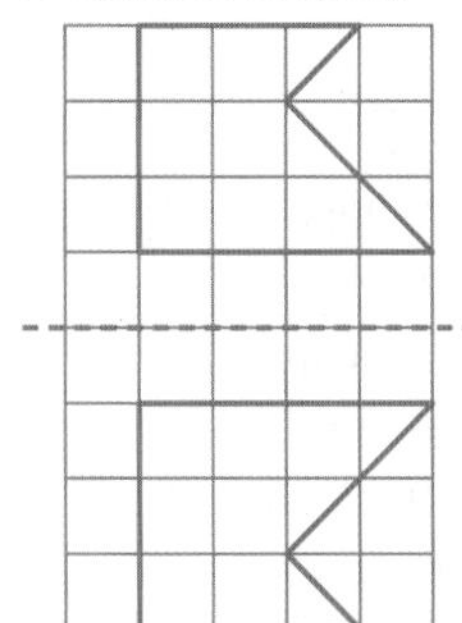

21 **a** $a = 40$ **b** $b = 104$
c $c = 20$ **d** $d = 115$
e $e = 40$ **f** $f = 70$

22 alternate angles

23 **a** Teacher to check **b** 7.1 cm

24

25 **a** rotation **b** UV
c $\angle W$

Challenge 27

Tests for congruent triangles PAGE 110

'... the conviction that we are loved.'

Congruent or different triangles? PAGE 112

1 congruent **2** different
3 congruent **4** impossible
5 congruent **6** congruent
7 different **8** congruent
9 impossible **10** congruent

Chapter 9

StartUp assignment 9 PAGE 120

1 **a** 512 **b** −32
2 66% **3** Yes
4 18.85 cm **5** $y - 4$
6 $\frac{3}{20}$ **7** $r = 70$
8 angles that add to 90° **9** $\frac{-20}{r}$
10 numerical **11** $\frac{2}{9}$
12 **13** $4xy(8y - 5)$

x

14 $p^2 = a^2 + d^2$

15 **a** $\frac{2}{5}$ **b** $\frac{5}{8}$
c $\frac{3}{4}$

16 **a** 21 **b** 3
c 4

17 **a** 90% **b** 12%

18 certain event

19 **a** $\frac{5}{8}$ **b** $\frac{2}{9}$
c 0.018

20 $\frac{2}{3}$ **21** $\frac{1}{4}$

22 56%

23 **a** $\frac{9}{25}$ **b** $\frac{13}{20}$

24 **a** $\frac{1}{4}$ **b** white

25 Teacher to check **26** red, amber, green

Challenge
a 24 **b** 24
c 8 **d** 8

Probability review PAGE 124

1 no chance at all **2** $\frac{4}{13}$

3 **a** $\frac{1}{4}$ **b** $\frac{5}{12}$
c $\frac{2}{3}$

4 **a** $\frac{1}{26}$ **b** $\frac{5}{12}$
c $\frac{2}{3}$ **d** $\frac{7}{13}$

5 $\frac{1}{2}$

6 **a** $\frac{3}{10}$ **b** $\frac{1}{5}$
c $\frac{13}{40}$

7 **a** $\frac{9}{20}$ **b** $\frac{11}{100}$
c $\frac{9}{50}$

8 $\frac{13}{20}$

9 **a** 4 **b** $\frac{1}{4}$
c $\frac{1}{2}$

10 $\frac{2}{125}$ **11** $\frac{1}{6}$

12 **a** $\frac{1}{5}$ **b** $\frac{3}{5}$
c $\frac{3}{5}$

13 **a** 12: H1, H2, H3, H4, H5, H6, T1, T2, T3, T4, T5, T6
b $\frac{1}{12}$ **c** $\frac{1}{4}$

14 a

		2nd die					
		1	2	3	4	5	6
1st die	1	0	1	2	3	4	5
	2	1	0	1	2	3	4
	3	2	1	0	1	2	3
	4	3	2	1	0	1	2
	5	4	3	2	1	0	1
	6	5	4	3	2	1	0

b $\frac{1}{6}$ **c** $5, \frac{1}{18}$

15 a 36% **b** $48\frac{1}{2}$%

16 a BBB, BBG, BGB, GBB, BGG, GBG, GGB, GGG

b $37\frac{1}{2}$%

17 $\frac{3}{7}$

Dice probability PAGE 126

1

		1st die					
	+	⚀	⚁	⚂	⚃	⚄	⚅
2nd die	⚀	2	3	4	5	6	7
	⚁	3	4	5	6	7	8
	⚂	4	5	6	7	8	9
	⚃	5	6	7	8	9	10
	⚄	6	7	8	9	10	11
	⚅	7	8	9	10	11	12

2

Sum	Probability	Probability (%)
2	$\frac{1}{36}$	2.78
3	$\frac{1}{18}$	5.56
4	$\frac{1}{12}$	8.33
5	$\frac{1}{9}$	11.11
6	$\frac{5}{36}$	13.89
7	$\frac{1}{6}$	16.67
8	$\frac{5}{36}$	13.89
9	$\frac{1}{9}$	11.11
10	$\frac{1}{12}$	8.33
11	$\frac{1}{18}$	5.56
12	$\frac{1}{18}$	2.78

3 7

4 6 and 8

5 to 8 Teacher to check.

Probability crossword PAGE 128

Across

1 half
4 heads
7 six
8 one
9 event
11 exclusive
13 fifty
15 probable
16 diamonds
17 two-way
18 certain
20 outcome
21 space
22 complementary
23 twelve

Down

2 frequency
3 unlikely
5 draw
6 sample
10 trial
12 impossible
14 random
19 probability

Chapter 10

StartUp assignment 10 PAGE 134

1 a 0.125 **b** 0.583

2 10 000

3 Not right–angled

4 a 5 **b** 12 **c** 2.2

5 0.65

6 130°

7 $\frac{c}{5}$

8 4^4

9 27°

10 37.5%

11 1

12 19.63 m^2

13 a 1 **b** 5

14 a 2 **b** 17 **c** 1

15 a $-2b + 8$ **b** $4h + 1$ **c** $4m + 3$ **d** $-2r - 4$ **e** $\frac{u}{4}+5$ **f** $2d$

16 a $2n + 3$ **b** $18 - x$

17 a -6 **b** 10

18 a -5 **b** 13 **c** -4 **d** 7 **e** 8

19 $\frac{3+7}{5}=\frac{10}{5}=2$

20 4

21 a $4n - 28$ **b** $5r + 5$ **c** $-6d - 18$

Challenge James 27, Elizabeth 15

Equations crossword PAGE 137

Across

3 check **8** pronumeral
9 variable **12** formula
16 translate **17** consecutive
9 solve **21** equation
23 unknown **24** guess

Down

1 difference **2** backtracking
4 product **5** perimeter
6 operation **7** sum
10 inverse **11** two-step
13 balancing **14** brackets
15 solution **18** undoing
20 expand **22** represent
25 substitute

Equation problems PAGE 138

1 3
2 Mark = \$7.40; Annabella = \$14
3 \$245 **4** 5
5 17
6 a 81 **b** 11
7 Length = 24, Width = 12
8 Banana = 24 g; Apple = 48 g; Orange = 52 g
9 a \$73.60 **b** 52 km
10 32, 33 **11** 7
12 9.6 m; 19.2 m, 19.2 m
13 Ronan = \$45; Chad = \$33; Atif = \$38
14 a 77°F **b** 38°C
15 a 7.4
b Length = 22.2 cm; Width = 17.8 cm

Chapter 11

StartUp assignment 11 PAGE 146

1 a $18 - 3p$ **b** $p = -4$
2 chord **3** 8
4 a \$2.72 **b** -32
5 $4m(11 - 2m)$ **6** Yes ($39^2 = 15^2 + 36^2$)
7 \$54 **8** $\frac{2d}{e}$
9 3.53 cm^2
10 a $\frac{1}{4}$
b Selecting a clubs, hearts or diamonds card
11 a equilateral **b** acute-angled
12 a $\frac{4}{9}$ **b** $\frac{9}{4}$
c 2 : 5
13 a 7 **b** 8
14 350
15 a $3\frac{1}{3}$ h **b** 6 h
c 40 min **d** 5 h 12 min
16 a 2900 **b** 0.4
c 85.1 **d** 7190

17 a 25 **b** 40
18 19 : 30 **19** 6 : 45 p.m.
20 a 18 **b** \$45
21 $\frac{1}{7}$
22 a 30 **b** 15
23 7 **24** 12 : 40 a.m.
Challenge A → B, A → C, B → C, A → B, C → A, C → B, A → B, A → C, B → C, B → A, C → A, B → C, A → B, A → C, B → C

Ratio applications PAGE 148

1 20 cm **2** \$18 288 000
3 10 faulty **4** 9 teachers
5 51 cm, 68 cm, 85 cm
6 a 5 kg
b The larger packet by 5 cents
7 20.4 tonnes **8** 1.2 m
9 49 cm by 35 cm
10 5 mL of B; 20 mL of glue altogether
11 Gaz \$50 000, Judy \$75 000, Kevin \$25 000
12 203.35 m
13 Sand 120 kg, Cement 30 kg
14 Length 2.85 m, Width 1.9 m
15 105 mL of Y; 150 mL of Z
16 a Petrol 136.8c/L **b** Petrol rises by 8c/L

Chapter 12

StartUp assignment 12 PAGE 160

1 \$22.35 **2** $\frac{9}{20}$
3 10 000
4

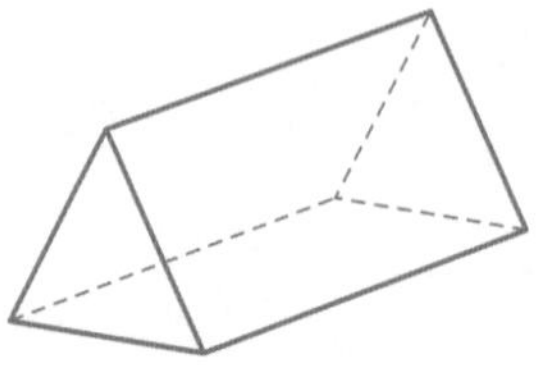

5 6.18 p.m.
6 Two angles that add to 180° **7** $\frac{1}{20}$
8 30 **9** 81
10 $x = 107$ **11** $30d^2e$
12 5 **13** $7p + 6$
14 \$213.20
15 A triangle with no equal sides.
16 a (0, 3) **b** (3, 1)
c $D(2, -1)$
17 a

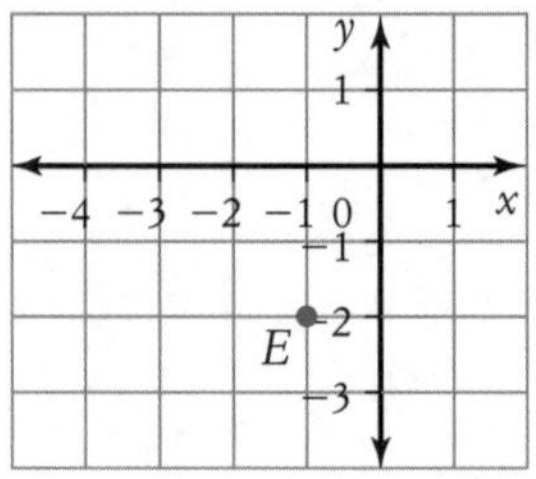

b 2nd
c D **d** B
18 a -3 **b** 22
c -7 **d** 27

19 16

20 **a** 2 **b** 8

21 (0, 0)

22 **a** $x = 4$ **b** $x = 9$

23 **a**

Number of tables	1	2	3	4
Number of chairs	6	10	14	18

b 34

24 **a** 2 **b** 3

c 5

25 **a** $(-3, -3), (2, 7)$ **b** $(0, 3)$

Challenge Mr Brown wore green, Mr Black wore brown, Mr Green wore black.

Finding the rule

PAGE 162

a $y = x + 7, 8, 7$
b $y = x - 4, 6, 0$
c $y = 4x, 32, 4$
d $t = \frac{h}{2}, 2, 4$
e $t = h + 8, 20, 15$
f $t = h - 3, 1, 5$
g $d = 6c, 60, 36$
h $d = c - 5, 11, 8$
i $d = c + 10, 19, 16$
j $p = 3k, 18, 21$
k $p = k + 8, 13, 14$
l $p = 2k + 2, 12, 14$
m $v = 3u - 4, 11, 14$
n $v = 5u + 1, 21, 26$
o $v = 2u - 4, 12, 14$
p $z = 4s + 2, 26, 30$
q $z = 3s + 3, 18, 21$
r $z = 8s + 2, 42, 50$
s $n = g + 4, 11, 12$
t $n = 3g + 9, 21, 24$
u $n = 7g, 35, 42$
v $m = 10i + 6, 66, 76$
w $m = 6i - 5, 25, 31$
x $m = 2i + 1, 3, 21$
y $m = 3i - 5, 7, 10$

Graphing linear equations

PAGE 164

1 $y = x + 3$

x	−1	0	1	2
y	2	3	4	5

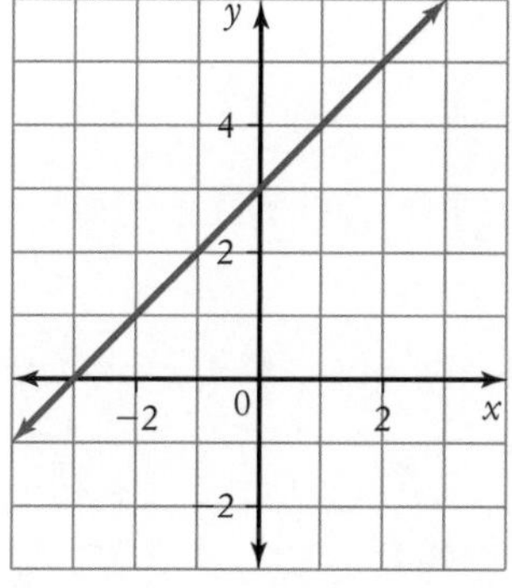

2 $y = 2x$

x	−1	0	1	2
y	−2	0	2	4

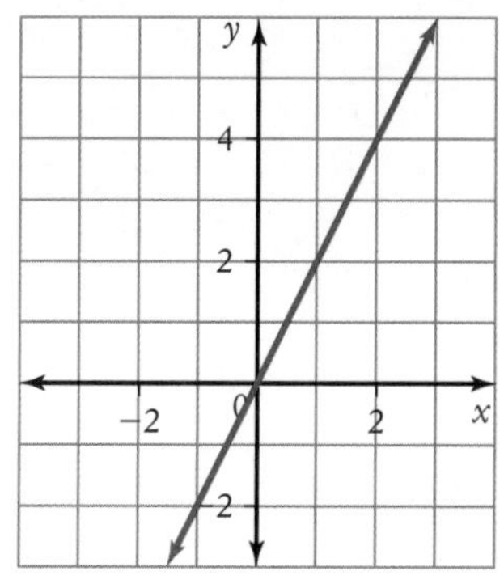

3 $y = \frac{1}{2}x + 1$

x	−1	0	1	2
y	$\frac{1}{2}$	1	$1\frac{1}{2}$	2

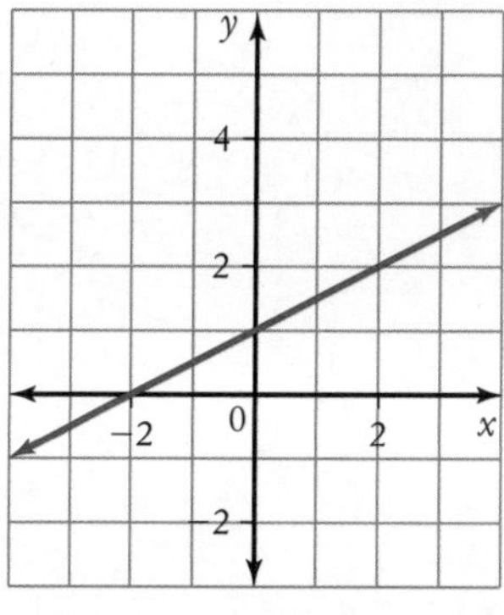

4 $y = 2x - 3$

x	−1	0	1	2
y	−5	−3	−1	1

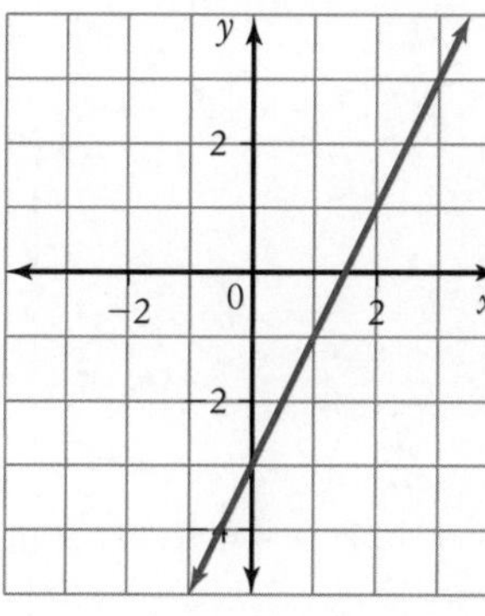

5 $y = 3x + 1$

x	−1	0	1	2
y	−2	1	4	7

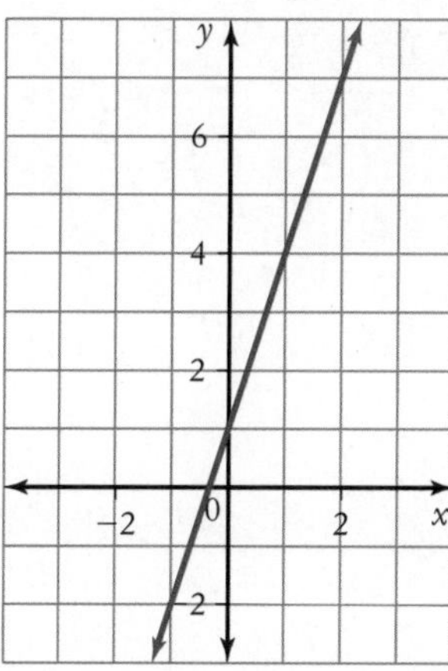

6 $y = -x + 5$

x	−1	0	1	2
y	6	5	4	3

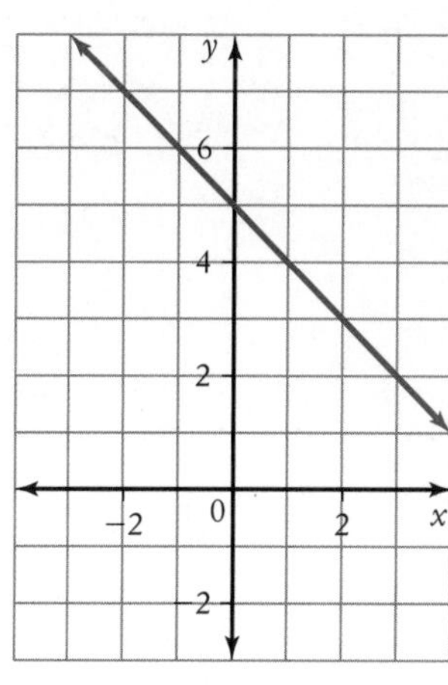

7 $y = -x - 1$

x	−1	0	1	2
y	0	−1	−2	−3

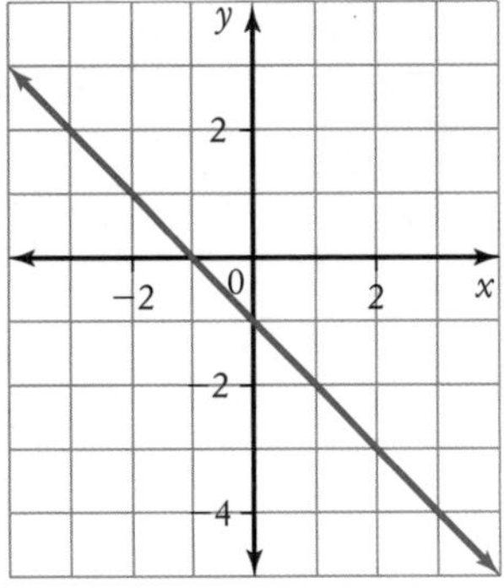

8 $y = -2x + 2$

x	−1	0	1	2
y	4	2	0	−2

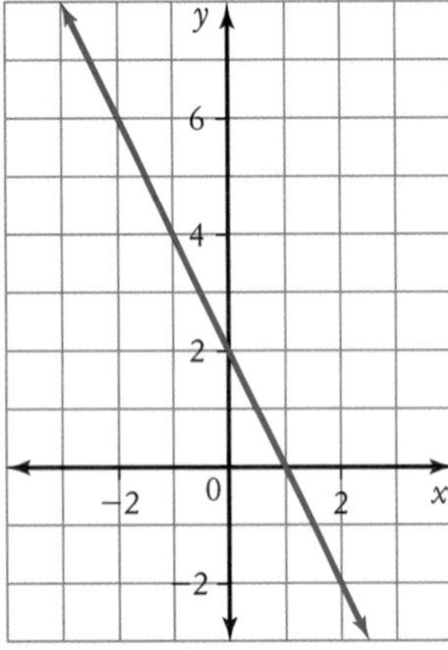

9 $y = -3x + 4$

x	−1	0	1	2
y	7	4	1	−2

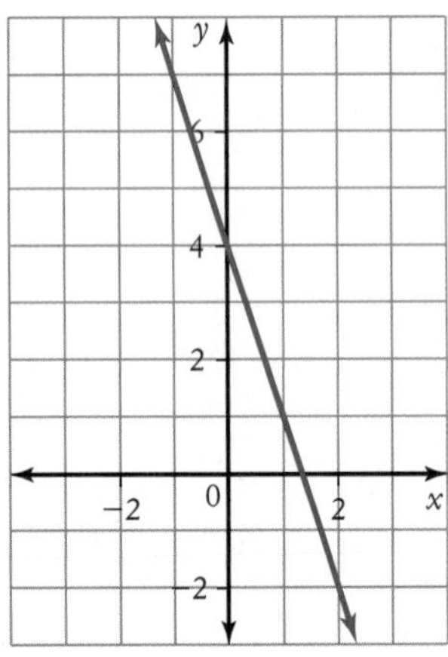

10 $y = x$

x	−1	0	1	2
y	−1	0	1	2

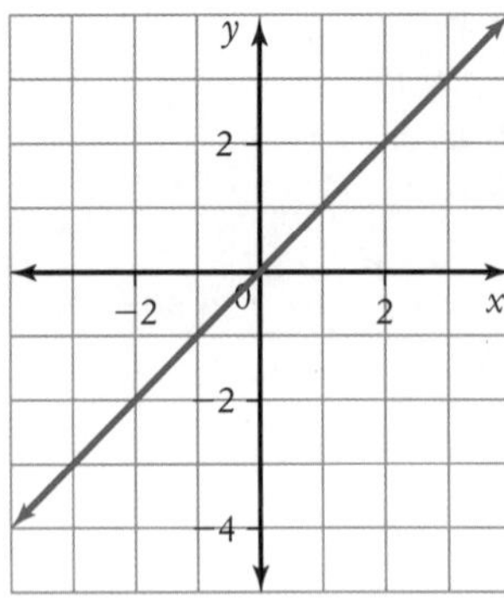

Linear equations crossword PAGE 172

Across

3 evaluate
7 substitute
8 plane
11 nonlinear
12 constant
13 axes
17 continuous
19 x-axis (or y-axis)
22 parallel
23 coefficient
25 y-axis (or x-axis)
26 pattern
29 slope
30 satisfy

Down

1 formula
2 rule
4 values
5 equation
6 decreasing
9 infinite
10 intersection
14 coordinates
15 linear
16 y-intercept
18 increasing
20 variable
21 steepness
24 arrows
27 table
28 graph

 9780170454520

Chapter 1

Pythagoras theorem
PAGE 09

Part A

1 9.08 p.m. **2** 608
3 50 cm^2 **4** 220
5 $4m(2 - m)$ **6** 76 cm
7 46 **8** $\frac{1}{14}$

Part B

1 4.12 **2** 73
3 a 9.75 **b** 11.18
4 $x = 175$ **5** 55
6 $y = 3$ **7** $m = 36$

Part C

1 a 17 m **b** $x = 15$
2 $\sqrt{3625}$ or $5\sqrt{145}$ **3** 14.1
4 $\sqrt{50}, \sqrt{33}, \sqrt{69}$ **5** 27.37
6 65 mm

Part D

1 a Length of the boundary of a shape, distance around the shape
b The longest side of a right-angled triangle
2 $c^2 = a^2 + b^2$ **3** surd
4 three squared
5
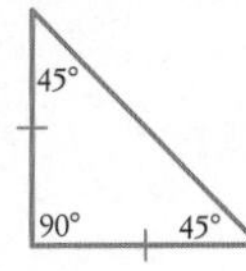

6 the sum of the square of the other 2 sides

Chapter 2

Integers
PAGE 16

Part A

1 a 8 **b** 528
2
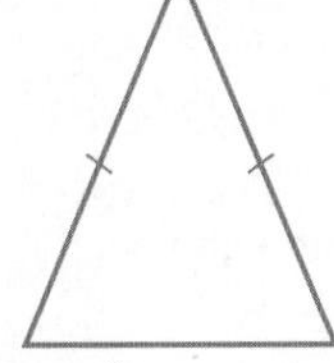
3 12
4 $x = 6$ **5 a** 22 cm
b 24 cm^2 **6** 360

Part B

1 a $>$ **b** $<$
c $<$ **d** $>$
2 8, 7, 0, −2, −4, −5, −7, −9 **3 a** 11
b −5

Part C

1 a 0 **b** −11
c 1 **d** −10
e 8 **f** 14
g −2 **h** 13

Part D

1 6.5, $\frac{1}{3}$ **2** 9°C
3 positive **4 a** 6°C
b 3°C **c** 0°C
5 Teacher to check: one or all of the integers must be negative, for example, −4, 5, 1
6 always positive

Decimals 1
PAGE 18

Part A

1 a 3 **b** $\frac{1}{3}$
2 $\frac{2}{3}$ **3** 22
4 Any answer of the form $(n, 0)$
5 $42\frac{1}{2}$ % **6** 3600
7
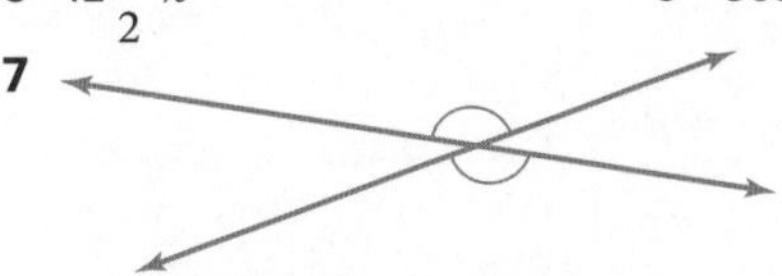
or mark the other 2 angles

Part B

1 1940 **2** $\frac{7}{25}$
3 4 **4 a** 20
b 46 **c** 36
5 0.03, 0.15, 0.2, 0.8, 1.5, 2.04

Part C

1 a 3.29 **b** 8.924
2 a 176.752 **b** 10.671
c 358.014 **d** 172.2
e 5.29 **f** 846.42

Part D

1 Teacher to check, for example, 16.543, 16.5361
2 a 1 **b** points, each
3 2.58 **4 a** $33
b $17

Decimals 2
PAGE 20

Part A

1 16 **2** 7
3 $36rv$ **4** $\frac{k}{7}$ or $k \div 7$
5 $d = 14$ **6** 12 000
7 $2\frac{2}{5}$
8 or any other shape with 6 straight sides

Part B

1 a 4.6 **b** 4.597

2 Teacher to check, for example, 7.043

3 $\frac{17}{20}$ **4 a** 11.73

b 0.025 **c** 6.844

d 578

Part C

1 6.041 **2 a** 841.2

b 210.3 **3** $15.\dot{1}3\dot{7}$

4 a 0.875 **b** $0.8\dot{3}$

5 a 0.00275 **b** 136

Part D

1 A decimal where one or a group of digits repeat endlessly, such as 5.272727 = $5.\dot{2}\dot{7}$

2 Teacher to check, for example, 2.254

3 Teacher to check, for example, 37.458 ÷ 0.9 = 374.58 ÷ 9 = 41.62, moving the point in both decimals one place to the right so that the second decimal is a whole number

4 a \$3.95 **b** 4.2 kg

5 a 4.9 **b** 4.9

Mental calculation and powers PAGE 22

Part A

1 31 **2** $\frac{3}{5}$

3 2 **4** 8

5 An angle less than 90° (smaller than a right angle) **6** 15 m^2

7 $\frac{7}{20}$ **8** bc

Part B

1 49 **2** 28

3 a 6^6 **b** 4^8

4 a 625 **b** 576

c 27 **d** 15

Part C

1 a 791 **b** 2400

c 765 **2 a** 64

b 7 **3** **a** 2^{11}

b 3^5 **c** 8^4

Part D

1 a 4 to the power of 5 **b** 5 (the power)

2 a \$14.70 **b** \$5.30

3 subtract

4 14 × 12 = 14 × 10 + 14 × 2 = 140 + 28 = 168

5 28

Powers and divisibility PAGE 24

Part A

1 4800 **2** $1\frac{2}{3}$

3 0.08 **4** $a + b + 7$

5

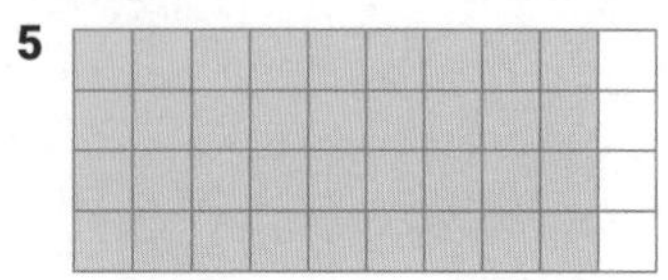

(or any 36 smaller squares shaded)

6 63 **7** reflex

8 13

Part B

1 a 81 **b** 49

c 8 **d** 4

2 a 5^{10} **b** 9^4

3 1, 2, 4, 5, 10, 20

Part C

1 a 3^6 **b** 5^{12}

2 a 1 **b** 1

c 1 **3 a** yes

b no **c** yes

Part D

1 a 25

b The number which if cubed gives 15 625; 25^3 = 15625

2 2, 3 **3 a** 63 513 (for example)

b The sum of the digits of the number is also divisible by 9; 6 + 3 + 5 + 1 + 3 = 18, which is divisible by 9.

4 a 4^0 **b** 1

c Equal to 1

Prime factors PAGE 26

Part A

1 a 125 **b** $\frac{2}{3}$

c 522 **d** 11.24

2 right angle **3** −6

4 7 **5** 50%

Part B

1 a 1, 2, 3, 6, 9, 18 **b** 1, 11

2 4, 8, 12, 16, 20, 24 **3** 23, 47, 61

4 4 **5** 60

6 12, 14, 15, 16, 18

Part C

1 a $72 = 2^3 \times 3^2$ **b** $120 = 2^3 \times 3 \times 5$

c $84 = 2^2 \times 3 \times 7$ **2 a** 24

b 504

Part D

1 a A number whose factors are 1 and itself, such as 3

b 2 **c** 1

2 highest common factor **3 a** 32

b 192

4 a 16, for example **b** square numbers

Powers and prime factors PAGE 28

Part A

1 121 **2** 156

3 An angle of size 180° **4** 8.3

5

(or any 8 smaller squares shaded)

6 4.40 p.m. **7** $\frac{3}{50}$

8 $2y - 10$

Part B

1 a 8^9 b 4^4

c 3^{15} d 2

2 2, 3, 5, 7, 11, 13, 17, 19

3 16 : 1, 2, 4, 8, 16; 32 : 1, 2, 4, 8, 16, 32; HCF = 16

Part C

1 −5

2 a $60 = 2^2 \times 3 \times 5$ b $140 = 2^2 \times 5 \times 7$

3 420

Part D

1 Multiply the prime factors of 60 by the non-common prime factors of 140: $2^2 \times 3 \times 5 \times 7 = 420$.

2 $10^4 = 10\,000$ 3 cube, root, 5, 125.

4 A number with more than 2 factors, such as 4.

5 multiply 6 14

Chapter 3

Algebraic notation and substitution

PAGE 36

Part A

1 a 512 b 12.32

2

3 38%

4 $n = 7$ 5 $38.4\ \text{m}^2$

6 4200 7 $a = 26°, b = 154°$

Part B

1 a 6 b −11

2 72 3 a $3b$

b $2b^2$ c 0

4 a −18 b 16

Part C

1 a $36y^2$ b n

c $16 + \frac{r}{2}$ 2 a $10 - a$

b $2b + 4$ c $\sqrt{\frac{c}{2}}$ or $\sqrt{\frac{1}{2}c}$

3 a 34 b 4

Part D

1 a $\frac{\$p}{4}$ b $14 - k$

2 a 5 times d, divided by 3 b −20

3 a Any number multiplied by 1 is itself

b Any number multiplied by 0 is 0

4 a 540° b $180 \times 3 - 360 = 180$

Simplifying algebraic expressions

PAGE 38

Part A

1 a 88 b 81.5

c 9 2 $n = \frac{1}{2}$

3 8 cm 4 15

5 40 6 {1, 2, 3, 4, 5, 6}

Part B

1 a $4n + 10$ b $n - 2$

c $5 - n^2$ 2 a $28dp$

b $6h$ c $\frac{16 + r}{2}$

3 a 4 b −13

Part C

1 a $2a + 2b$ b $11w - 5$

c $-a + 5b$ d $7mn - 8m^2$

e $15bc$ f $10m^2$

g $3a$ h $-\frac{4m}{n}$

Part D

1 Terms that have exactly the same variable(s), such as $4d$ and $9d$, or $2pq$ and qp

2 a Any number added to itself equals 2 times that number

b Any number divided by itself equals 1

3 Teacher to check, for example, if $a = 3$,
LHS = $5a - 4a = 5 \times 3 - 4 \times 3 = 3$
RHS = $a = 3$
So LHS = RHS and $5a - 4a = a$.

4 a $32y^2$ b $12y$

c 2 d $4y$

Expanding and factorising expressions

PAGE 40

Part A

1 40 2 a 49

b 0.021 3

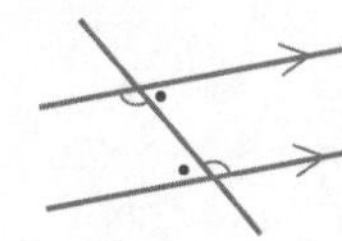

4 8 : 3 5 −17

6 $\frac{1}{8}$ 7 $20.5\ \text{m}^2$

Part B

1 a $5d - 13$ b $4r + 4$

c $24n$ d $-25c$

2 a 40 b 40

3 a 6 b 7

Part C

1 a $5m - 20$ b $-7a - 42$

c $8a - 12$ 2 a $3y$

b $4a$ 3 a $4a(3 + 4b)$

b $2b(5 - 2b)$ c $-9n(2m + p)$

Part D

1 a $m + 31$ b $2e^2 + 22e$

2 5 is a factor but not the highest common factor because $(4a - 3ab)$ still has a common factor of a; the correct factorisation is $5a(4 - 3b)$.

3 a $ab + ac$

b $27 \times 11 = 27 \times (10 + 1) = 27 \times 10 + 27 \times 1 = 270 + 27 = 297$

4 highest, factor 5 $8s(t + 3r - 2v)$

6 When $m = 5$, LHS = $7 \times 5 - 14 = 21$, RHS = $7 \times (5 - 2) = 21$, so LHS = RHS and $7m - 14 = 7(m - 2)$

HW HOMEWORK ANSWERS

Algebra review PAGE 42

Part A

1 148 **2** $\frac{5}{8}$

3

4 06:05

5 a 37.5% **b** $\frac{3}{8}$

6 $n = 6$ **7** 9 m

Part B

1 a $2w + 3v$ **b** $13a$

c $-30stv$ **d** $\frac{p}{4} - r^2$

2 a $9w + 45$ **b** $3mn - 2m$

3 a $3x - 5y$ **b** $\frac{x+y}{2}$

Part C

1 $2w + 2y$ **2 a** -23

b 51 **3** $3a$

4 a $12b(ac + 4d)$ **b** $-5a(3w + 2b)$

5 a $-12f$ **b** $16m - 19$

Part D

1 a A number, n, with 4 added, then all multiplied by 3

b $3(n + 4) = n + 4 + n + 4 + n + 4 = 3n + 12$

2 Teacher to check, for example, if $z = 5$, LHS $= 7 \times 5 - 5 \times 5 = 10$, RHS $= 2 \times 5 = 10$, LHS = RHS and $7z - 5z = 2z$.

3 $\frac{18a^2d}{2d} = 9a^2$ **4 a** $ab - ac$

b $16 \times 9 = 16 \times (10 - 1) = 16 \times 10 - 16 \times 1 = 160 - 16 = 144$

5 64.4°F

Chapter 4

Angle geometry PAGE 56

Part A

1 $w = 7\frac{1}{2}$ **2** 6.35 p.m.

3 5 : 4 **4** 4.6

5 $\frac{5}{7}$ **6** 24 m

7 1.6 **8** $27mn$

Part B

1 a acute **b** straight

c reflex **d** right

2 a

b

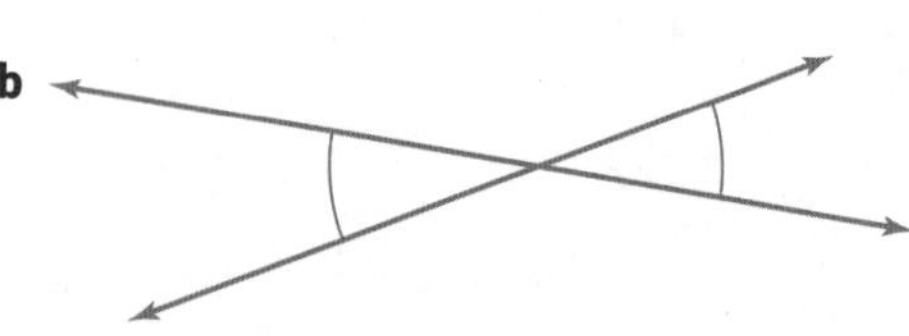

or mark the other 2 angles

3 a 132 **b** 90, 180

Part C

1 a $n = 47$ **b** $m = 23$

c $x = 152$ **d** $v = 74°$

2 $x = 22$ (angles on a straight line), $y = 158$ (corresponding or co-interior angles on parallel lines)

Part D

1 a 360 **b** 180

2 adjacent

3 a Corresponding angles are equal.

b 

c

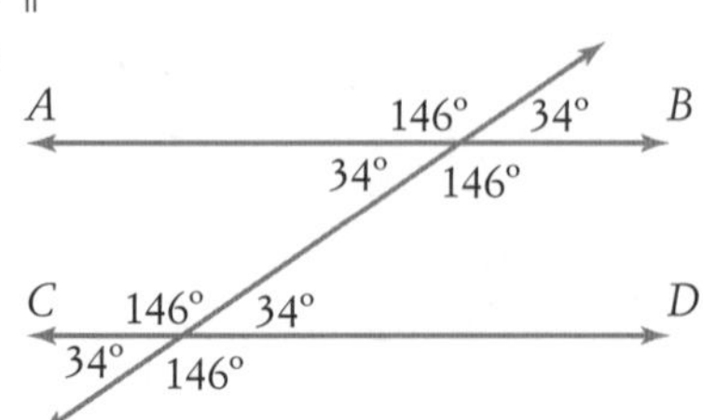

4 a equal **b** co-interior

Symmetry and triangles PAGE 58

Part A

1 64 **2** 43 min

3 Teacher to check, for example, 15 : 24

4 a 216 **b** -150

5 42 m^2 **6** $\frac{b+k+x}{3}$

7 $6a^2m$

Part B

1 a isosceles **b** equilateral

2 a $n = 115$ (angles on a straight line)

b $m = 147$ (vertically opposite angles)

3 a

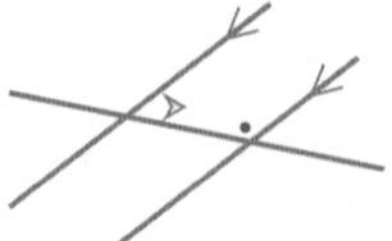

or

b are supplementary (or add up to 180°)

Part C

1 a

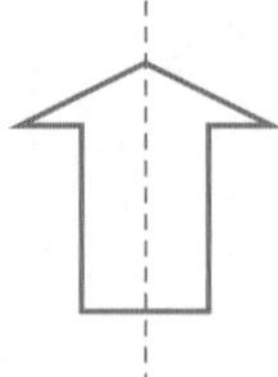

b

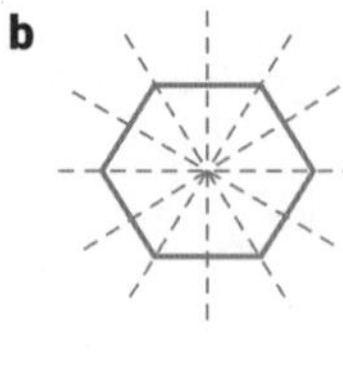

2 a 4 **b** 8

3 60° **4 a** scalene

b right-angled **5**

Part D

1 a A triangle with 2 equal sides

b 1

c no rotational symmetry

d 2 **2 a** scalene

b equilateral **c** obtuse-angled

3 In a right-angled triangle, the side opposite the right angle (the hypotenuse) is always longer than the other 2 sides

9780170454520

HW HOMEWORK ANSWERS

Quadrilaterals and angle sums PAGE 60

Part A

1 $d - 8$ **2** 96.3
3 a 16 **b** 0.05
4 $n = -4$ **5** 2150
6 $1\frac{1}{2}$ **7** 54 cm^2

Part B

1 (triangle) **2** rhombus
3 a i 2 **ii** 2
b i 3 **ii** 3
4 a equilateral **b** acute-angled

Part C

1 a trapezium **b** kite
2 a 90° (or equal) **b** parallel sides
3 a $r = 50$ **b** $x = 101$
c $m = 115$ **d** $n = 64$

Part D

1 a 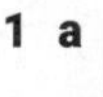

b Diagonals are of equal length and bisect each other
2 a A quadrilateral with 4 equal sides
b A square has 4 angles of size 90°
c 108°, 72°, 108°
3 Yes, because opposite sides are parallel (the definition of a parallelogram)
4 a 180° **b** $180 \div 3 = 60°$

Chapter 5

Area 1 PAGE 68

Part A

1 $\frac{2}{5}$ **2** 2208
3 5 : 4 **4** 16
5 8 **6** 120 m^3
7 $240 **8** $130.76

Part B

1 a 140 **b** 3.9
c 72 500 **2 a** 77 m^2
b 36 m **3 a** trapezium
b parallelogram **4** 6 m

Part C

1 a 21 m **b** 16.8 cm
c 17 m **2 a** 7000
b 4.52 **3 a** 10.5 cm^2
b 10.88 m^2 **4** $A = bh$

Part D

1 The distance around a shape, the total length of its sides.
2 square centimetre
3 hectare **4** 30 cm
5 Area of a triangle, A = area, b = base length, h = perpendicular height
6 0.3 m^2 **7** 13.5 cm

Area 2 PAGE 70

Part A

1 5.48 p.m. **2 a** $36
b 200 **3** 1
4 $\frac{7}{13}$ **5** 25 000
6 12 **7** 27 m^3

Part B

1 a 10.24 m^2 **b** 12.8 m
2 a 25.5 m^2 **b** 1.08 m^2
3 a 90 000 **b** 1 300 000
4 any 2 numbers with a product of 30, such as 5 and 6
5 4 cm

Part C

1 a 27 m^2 **b** 24 m
2 a 17.28 m^2 **b** 23.46 cm^2
c 42 m^2 **d** 20 m^2
e 1395 mm^2 **f** 33 m^2

Part D

1 a Area of a trapezium
b A = area, a and b are the lengths of the parallel sides, h = perpendicular height
2 a 2795 cm^2 **b** 0.2795 m^2
3 a

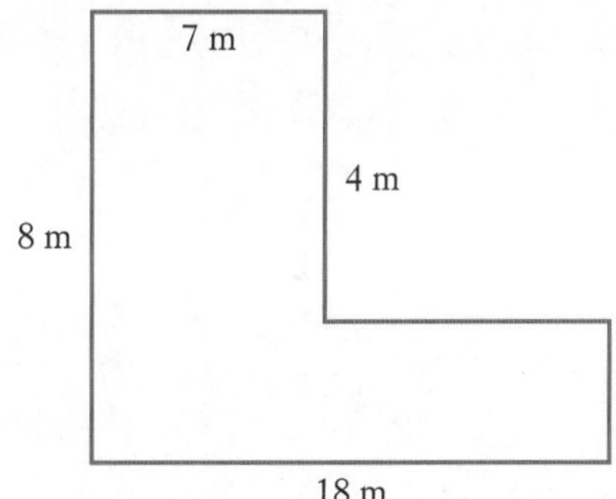

b 52 m **c** 100 m^2
4 Find half of the product of the lengths of the diagonals of the rhombus.

Circumference and area of a circle PAGE 72

Part A

1 -125 **2** $2mn + 9$
3 $\frac{7}{11}$ **4** 5 h 40 min
5 0.6 **6 a** scalene
b right-angled **7** 7.7

Part B

1 16 m^2 **2 a** 56 000
b 4.685 **3 a** 37.2 m^2
b 30.38 m^2 **4**

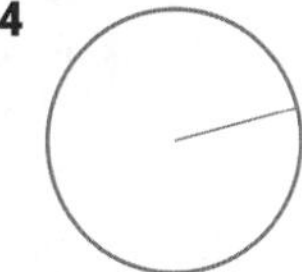

5 a 27 cm **b** 26 m

Part C

1 a 25.13 m
b 18.85 m
c 25.71 cm
d 14.28 cm

2 a 28.27 m^2
b 153.94 cm^2
c 31.81 cm^2
d 19.63 m^2

Part D

1 irrational

2 The perimeter or distance around a circle

3 a 
b 16.59 cm^2
c 1

4 $A = \pi r^2$

5 a 9.42 m
b 7.07 m^2

Volume and capacity

PAGE 74

Part A

1 80%

2 1805

3 a $\frac{4}{15}$
b 13.1

4 $-18ad$

5 4.5

6 a 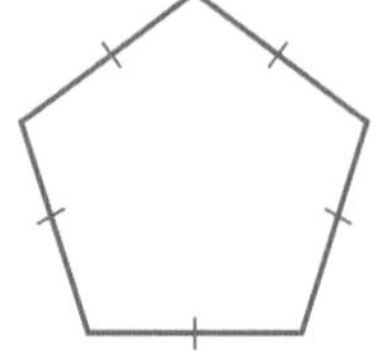
b 20 m

Part B

1 a 12 000
b 390
c 10

2 a 72.9 cm^2
b 19 m^2
c 60 m^2

3 27 m^3

4 19.63 m^2

Part C

1 a 3 200 000
b 108

2 a 288 m^3
b 42 cm^3
c 96 m^3

3 a circle
b 3.08 m^3
c 3079 L

Part D

1 The amount of space taken up by an object

2 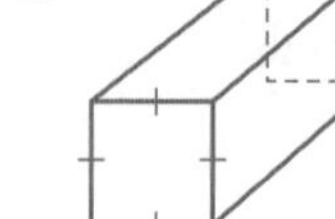

3 1000

4 a 280 cm^3
b No

5 a $V = \pi r^2 h$
b V = volume, r = radius, h = perpendicular height

6 707 mL

Chapter 6

Fractions

PAGE 84

Part A

1 64

2 2 equal sides

3 Teacher to check, for example, 3.547

4 5

5 a $-5a$
b k

6 A plane shape with 8 straight sides

7 $m + n - 5$

Part B

1 a $\frac{5}{6}$
b $\frac{2}{5}$

2 a $3\frac{2}{7}$
b $1\frac{1}{3}$

3 $\frac{5}{6}$

4 8

5 $\frac{23}{5}$

6 $\frac{1}{2}$

Part C

1 a $1\frac{7}{15}$
b $\frac{1}{8}$
c $4\frac{2}{15}$
d $1\frac{1}{3}$
e $\frac{4}{5}$
f $1\frac{1}{2}$
g $8\frac{1}{20}$
h $\frac{3}{4}$

Part D

1 a $2\frac{2}{3}$
b The answer is 1

2 a $\frac{3}{40}$
b 40

3 a denominator
b multiply

4 a $\frac{4}{9}$
b $\frac{5}{8}$

Percentages 1

PAGE 86

Part A

1 a 5
b 8

2 4 and 5

3 $60x$

4 7.5 cm

5 n^3

6 360°

7 105°, 75°, 105°

Part B

1 a $\frac{7}{25}$
b $\frac{11}{20}$

2 a 0.07
b 0.185

3 2 h 40 min

4 0.4, 65%, $\frac{3}{4}$, $\frac{7}{8}$

5 a 88%
b 64%

Part C

1 a $385
b 960 mL

2 a $\frac{3}{8}$
b 37.5%

3 $975.20

4 a 7%
b 35%

5 $216.75

Part D

1 a $\frac{2}{3} \times 100\% = 66\frac{2}{3}\%$
b $70\% = \frac{70}{100} = \frac{7}{10}$

2 a add
b 140
c 60

3 Convert 5 cm to 50 mm: $\frac{45\text{ mm}}{50\text{ mm}} \times 100\% = 90\%$

4 22

Percentages 2

PAGE 88

Part A

1 a 24
b 0.054

2 $w = -28$

3 a 8
b 4.7

4 5 m

9780170454520

5 a

or

b equal (in size)

Part B

1 75%
2 $12
3 96
4 a $\frac{2}{5}$
b $\frac{1}{8}$
5 90%
6 $25
7 $432

Part C

1 $162.80
2 a $180
b $35
c 19.4%
3 $600
4 a $2200
b 13.75%
5 $62.50

Part D

1 Goods and services tax
2 $10 575
3 a loss
b profit
4 77.5%
5 a cost price, 100%
b denominator, 100
6 $82 600

Fractions and percentages revision PAGE 90

Part A

1 4
2 4200
3 a 21
b 4.85
c 1
4 $\frac{3}{13}$
5 $-44mn$
6 21

Part B

1 $2\frac{2}{3}$
2 75%, $0.\dot{7}$, $\frac{4}{5}$, $\frac{7}{8}$
3 a $\frac{7}{25}$
b $16\frac{2}{3}\%$
c 0.123
4 62.5%
5 a $1\frac{1}{2}$
b $3\frac{7}{15}$

Part C

1 76%
2 a $1\frac{2}{9}$
b $\frac{2}{15}$
3 $15.33
4 a 15 h
b 112.5 cm
5 a $80
b $5.60

Part D

1 a cost price, selling price
b 87
2 a a fraction whose numerator is greater than (or equal to) its denominator, such as $\frac{9}{7}$
b mixed numeral
3 a $560
b 680
c 21%
4 multiply by 100%

Chapter 7

Statistics 1 PAGE 100

Part A

1 8.235
2 4.55 p.m.
3 $5ab - 5b$
4 183.50
5 $\frac{13}{18}$
6 4
7 $\frac{3}{8}$
8 25°, 25°

Part B

1 a 7
b 3
c 2
2 a 5, 5, 5, 7, 8, 12, 12, 14, 20
b 20
c 5
d 8
3 120°

Part C

1 a 11
b 11
c 16.8
d 11
e 13
f 16
2 a 8
b 6.5

Part D

1 a mean
b mode
2 difference, highest, lowest
3 a 9
b 10
c 5
4 a 1
b The middle value when the values are arranged in order.

Statistics 2 PAGE 102

Part A

1 30
2 $m = 3\frac{2}{3}$
3 None
4 a 48
b $\frac{7}{12}$
5 a triangular prism
b 18 m^3
6 $\frac{2}{3}$

Part B

1 a 8
b 6.5
c 7
d

2 3 4 5 6 7 8 9

2 a 4
b 11
c 12.7
d 46

Part C

1 a Total frequency = 16, fx column: 3, 8, 6, 24, 5, 46
b 2.875
c 4
2 a 7 and 10
b 8
3 a 23
b 18
c 24

Part D

1 a 2

b The average of the 2 middle values when the values are arranged in order

2 mean, adding, dividing

3 a

Stem	Leaf
6	1 4 6
7	0 1 4 4
8	5 8
9	2
10	
11	
12	5 8

b 83.17

c 67

d 74

Statistics 3 PAGE 104

Part A

1 $0.\dot{4}$

2 $24a + 40$

3 360°

4 a 5

b $\frac{9}{40}$

5 $\frac{2e}{5d}$

6 37.5 m^2

7 $\frac{1}{4}$

Part B

1 a 7

b 2

c 2

d 2.6

2 a 4

b 3

3 a 24

b 23.5

Part C

1 a biased

b random

c biased

2 a sample

b sample

3 a Melbourne

b Melbourne

c Melbourne

Part D

1 a outlier

b census

2 chance

3 Teacher to check

4 a mean

b mode

5 Teacher to check. For example, not everyone shops, and not only on Saturday afternoon.

Chapter 8

Congruent figures 1 PAGE 114

Part A

1 a 45 000

b 1 000 000

2 $-8a + 12$

3 84.21

4 $60

5 32 m^2

6 4 h 45 min

7 32

Part B

1 a scalene, right-angled

2 a translation

b reflection

3 a

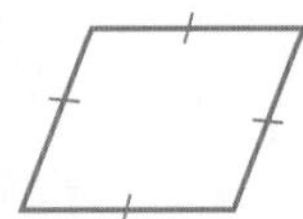

b 2

c bisect each other, at right angles or bisect the angles of the rhombus

4

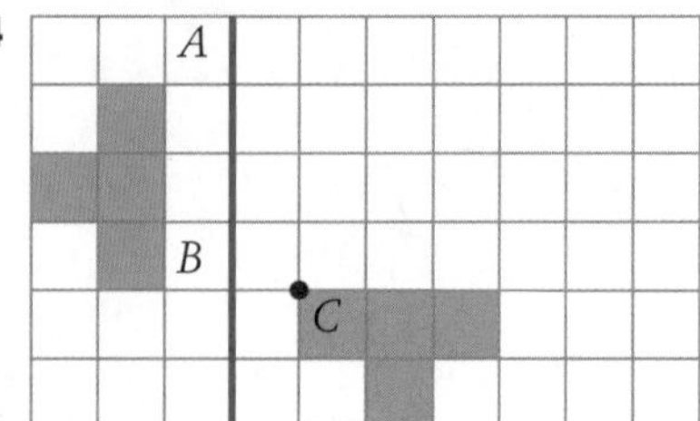

Part C

1 a rotation

b JM

c $\angle L$

d $KLMJ$

2 A and E, C and F

3 a X and Z

b RHS

Part D

1 They are identical in size and shape

2 a included angle

b right angle

c hypotenuse

3 'is congruent to'

4 SAS

5 a

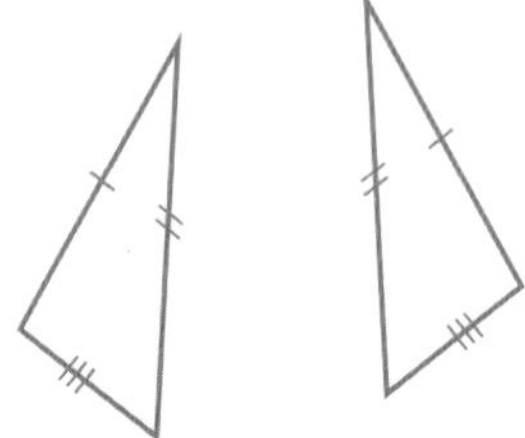

b

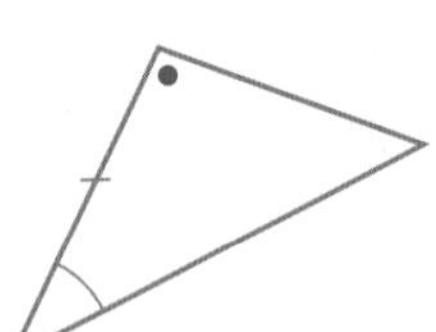

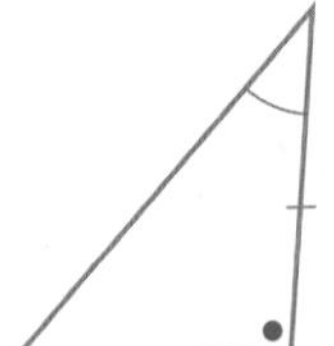

Congruent figures 2 PAGE 116

Part A

1 $\frac{1}{3}$

2 10 000

3 4 h 50 min

4 64°

5 a 10.8

b $1\frac{1}{5}$

6 $6(3a + 4)$

7 48 cm^3

Part B

1 a trapezium

b rhombus

2 a TR

b $\angle S$

c WVU

3 a A quadrilateral with all angles 90°

b 2

c Equal or bisect one another.

Part C

1 Teacher to check

2 a SSS

b all 90°

9780170454520

c

d No **e** bisect

Part D

1 Lines that cross at right angles / 90°

2 SAS, RHS

3 square, kite, rhombus

4 A quadrilateral with 2 pairs of parallel sides

5 AAS

6 **a** rectangle, square **b** 90°

Congruent figures revision

PAGE 118

Part A

1 **a** 6 000 000 **b** 3800

2 11.45 a.m. **3** $5ab$

4 **a** $\frac{7}{8}$ **b** 16.91

5 54 m^2 **6** $42

Part B

1 **a** SAS **b** QR

c $\angle X$ **d** YZX

2 **a** A quadrilateral with 4 equal sides and 4 right angles

b 4 **c** Yes

d bisect, right angles / 90°

Part C

1 Teacher to check **2** **a** SAS

b matching angles in congruent triangles

c 90° **d** $\angle L$

e angles, opposite

Part D

1 square, rectangle, rhombus, parallelogram

2 square, rhombus

3 **a** SSS **b** $\angle CAD$

c 65°

4 **a**

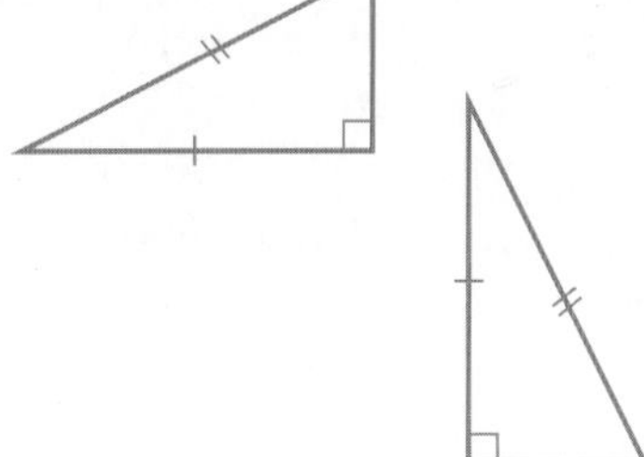

b 

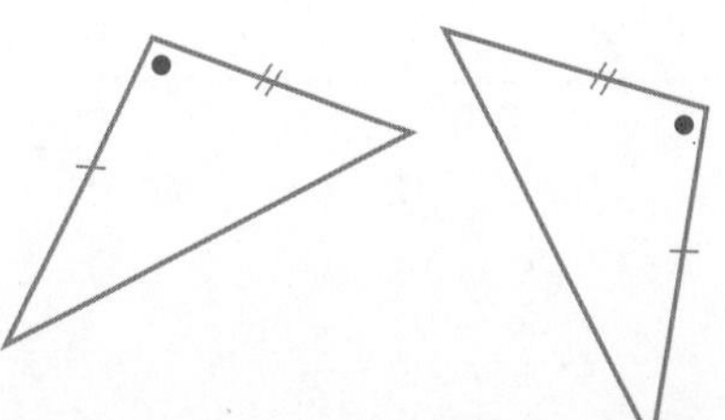

Chapter 9

Probability 1

PAGE 130

Part A

1 **a** 3.286 **b** 24 000

2 $423 **3** 5

4 $-70ab^2$ **5** $0.\dot{3}$

6 21.8 **7** $f = 120$

Part B

1 **a** 0.425 **b** 42.5%

2 **a** France wins, South Africa wins, a drawn match

3 1 **4** **a** unlikely

b certain **c** likely

d impossible

Part C

1 **a** 6 **b** 9

2 0.5 **3** 81%

4 **a** $\frac{2}{3}$ **b** $\frac{1}{3}$

5 **a** $\frac{17}{30}$ **b** $\frac{2}{15}$

Part D

1 **a** male, female

b can be male or female, but not both

2 **a** $\frac{11}{12}$ **b** $\frac{1}{2}$

3 **a** being born in a month not beginning with A

b $\frac{5}{6}$

4 **a**

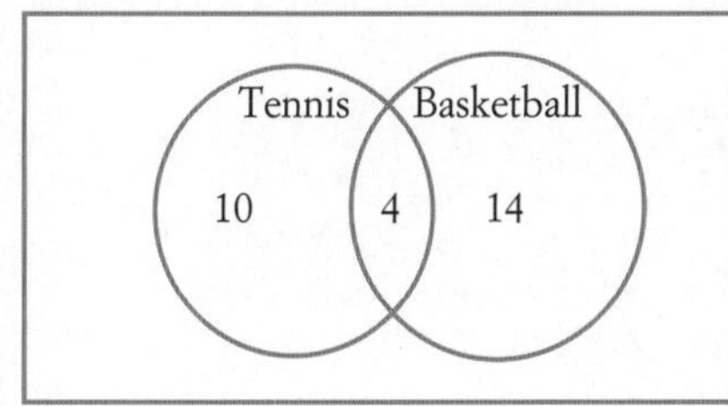

b $\frac{1}{7}$

Probability 2 PAGE 132

Part A

1 72
2 a $8a - 6b$
b $\frac{2}{3p}$
3 3
4 163.0
5 $\frac{5}{6}$
6 8 h 20 min
7 45°

Part B

1 a $\frac{1}{2}$
b $\frac{1}{2}$
2 1
3 a $\frac{11}{15}$
b 1
4 a The train arriving late
b 0.4
5 Each of those 3 outcomes is not equally likely

Part C

1 a 220
b 75
c i $\frac{6}{11}$
ii $\frac{8}{55}$
2 a $\frac{7}{20}$
b 140
3 a

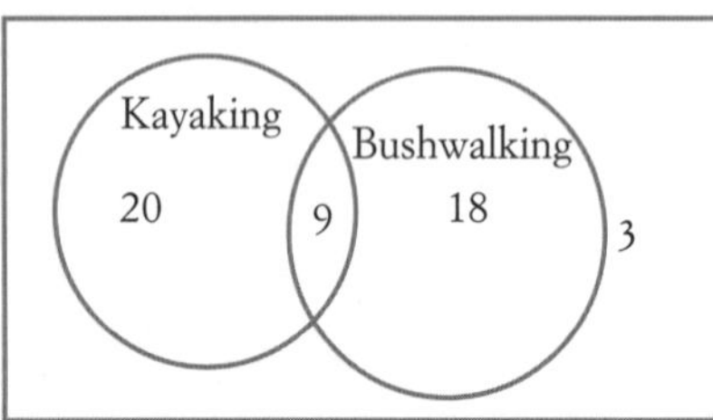

b $\frac{27}{50}$

Part D

1 a 0
b impossible
2 a a low probability or improbable
b 65%
c 95%
3 a 116
b i 0.3375
ii 0.8225

Chapter 10

Equations 1 PAGE 140

Part A

1 1, 2, 4, 8, 16, 32
2 210
3 0.65
4 $7ab - 2$
5 kite

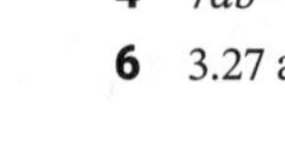

6 3.27 a.m.
7 32 m^2

Part B

1 a -19
b 22
2 a subtracting 4
b multiplying by 5
c adding 6
3 a -10
b 10
4 $6 \times 4 + 9 = 33$

Part C

1 a $m = 2$
b $x = 6$
c $y = 5\frac{1}{4}$
d $a = 6$
e $m = 3\frac{2}{3}$
f $r = -8$
g $x = 36$
h $a = 11$

Part D

1 solution
2 a $12 - 4n$
b $n = -4$
3a The sum of k and 9, all divided by 2, is equal to 10.
b $k = 11$
c $\frac{11+9}{2} = 10$
4 a $2m + 8 = 15$
b $m = 3\frac{1}{2}$

Equations 2 PAGE 142

Part A

1 a $\frac{11}{20}$
2 0.24
3 $w = 118$
4 60%
5 7 : 4
6 30 m
7 24
8 12 cm^2

Part B

1 a $4k - 28$
b $10t + 22$
2 a $7m - 2$
b $4c - 16$
3 a $b = -11$
b $m = -8$
c $w = -4$
d $m = 18$

Part C

1 a $d = -11$
b LHS $= 2 \times (-11) + 3 = -19$, RHS $= -11 - 8 = -19$, LHS = RHS
2 $y = 6$
3 a $x + x - 4 + x - 4 = 43$ or $3x - 8 = 43$
b Katie is 17

Part D

1 Teacher to check, for example, $y + 6 = 5$
2 variable, pronumeral or unknown
3 Teacher to check
4 Teacher to check, for example, $3p - 2 = 19$
5 a 3
b $3 \times 3 + 7 = 2 \times 3 + 10 = 16$

Equations revision PAGE 144

Part A

1 32
2 18 : 34
3 35%
4 $c = 32$
5 0.4
6 10
7 $-2(d - 7)$
8 300 m^3

Part B

1 LHS $= 5 \times (-16) + 4 = -76$, RHS $= 3 \times (-16) - 28 = -76$, LHS = RHS
2 LHS $= 6 \times (2 \times 6 - 5) = 42 =$ RHS
3 a $w = -8$
b $b = 5$
c $m = 60$
d $e = 15$
e $w = 3$
f $a = 2$

Part C

1 13
2 a $x = -4$
b $a = -5$
3 38
4 a \$141
b 2.5 h

Part D

1 a 16, 17, 18
b $y + 1, y + 2, y + 3$
2 a $x = 3$
b 24 cm
3 a $x + x + 1 + x + 2 = 156$ or $3x + 3 = 156$
b 51, 52, 53
4 a 86°F
b 10°C

9780170454520

Chapter 11

Ratios 1

PAGE 152

Part A

1 25

2 a $\frac{3}{25}$

b 0.12

3 $a = 6\frac{1}{2}$

4 30 m^2

5 $-60ab$

6

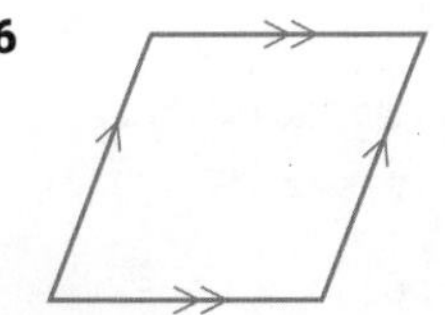

7 3.23

Part B

1 a $\frac{2}{5}$ **b** 3

2 36

3 9 h

4 a 20 **b** 3600 **c** 4400

5 320

Part C

1 a 20 **b** 8

2 20 kg

3 a 13 : 7 **b** 1 : 30 **c** 8 : 15 **d** 2 : 25

4 $396

Part D

1 a 7 : 3 **b** 3 : 10

2 a for every 1000 people in Australia there are 12 new births

b 300 000

3 a common factor (or common divisor)

b power

4 a 945

b 44

Ratios 2

PAGE 154

Part A

1

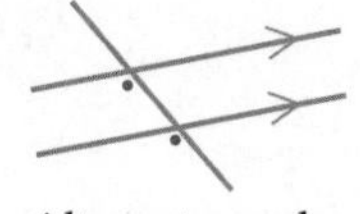

or

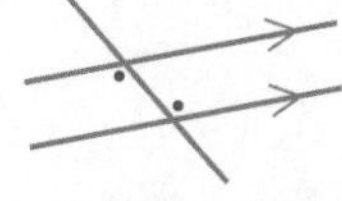

Alternate angles on parallel lines are equal

2 4

3 $\frac{1}{8}$

4 $6\frac{2}{5}$

5 36.8 cm^2

6 19

7 112°, 68°, 112°

8 2

Part B

1 a 3 **b** 45

2 a 65 **b** 143

3 a 3 : 7 **b** 3 : 1 **c** 1 : 13 **d** 1 : 18

Part C

1 a 10 kg **b** 70 kg

2 a 3 : 200 000 **b** 11 km

3 $450, $270

4 16

5 a 6000 **b** 33 000

Part D

1 a Distances on the plan represent actual distances that are 500 times bigger

b Distances on the plan represent actual distances that are 25 times smaller

2 Calculate the total number of parts by adding: 8 + 3 = 11

3 a 4 : 3 : 1

b Nadine $28 000, Janine $21 000, Renee $7000

c Add them to check that they add up to the total prize: $28 000 + $21 000 + $7000 = $56 000

4 Jackson's drink is sweeter, because it has fewer parts of water (1 : 6) than Stefan's (1 : 8), so it has a higher proportion of cordial.

Rates

PAGE 156

Part A

1 a 15 **b** 0.116

2 $a = 6\frac{1}{2}$

3 31.4 cm

4 0.32

5 −10

6 rhombus

7 add up to 90°

Part B

1 1650 c

2 a 60 words/min **b** 110 km/h **c** $17/h

3 a 8100 **b** 3250

4 $17.75

5 850

Part C

1 92 km/h

2 a 1290 **b** 23 min

3 C

4 $67.14

5 5 h 52 min

6 32 h

7 315 km

Part D

1 Average speed = $\frac{\text{distance}}{\text{time}}$ or $\frac{d}{t}$

2 a $6\frac{1}{2}$ hours **b** 3.15 p.m.

3 a kilometres per hour

b At this speed, in one hour the car will travel 85 km

4 unit, lowest/cheapest

5 a 2.25 goals/match

b Zac, as his rate was slightly higher at 2.5 goals/match

Travel graphs and time

PAGE 158

Part A

1

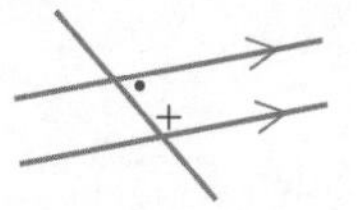

or

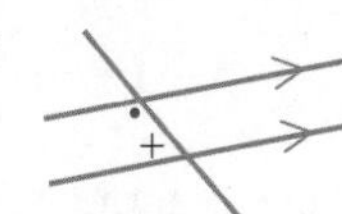

Co-interior angles on parallel lines are supplementary (add up to 180°)

2 0.625

3 a 343 **b** 4.05

4 12.56 m^2

5 5^3

6 $r + 54$

Part B

1 16 : 25

2 7 : 50 p.m.

3 1, 40

4 8 h

9780170454520

5 a 90 km/h **b** 3, 20
6 11 : 55 a.m. **7** C

Part C

1 a 8 : 30 a.m. **b** Grace stopped
c 12 p.m. **d** 1 h
e 10 km/h **2 a** 19 min
b Bobtown **c** 10 : 11

Part D

1 $20 > 12$, so it's p.m. time. $20 - 12 = 8$ p.m.
The time is 8:35 p.m.
2 a 836 km **b** 7 : 45 p.m.
3 a The person/car is stationary, stopped
b Returning to home/start
4 Jabira $18 600, Lily $11 160

Chapter 12

Tables of values PAGE 166

Part A

1 1, 2, 4, 5, 10, 20 **2** $45
3 $3b(3a + 4)$ **4 a** isosceles
b 17.6 m **c** 4 m
5 $0.8\dot{3}$ **6** $m = -8$

Part B

1 a 31 **b** 3
2 -3 **3 a** 5
b 3 **4 a** (0, 0)
b x-axis **c** 4th

Part C

1
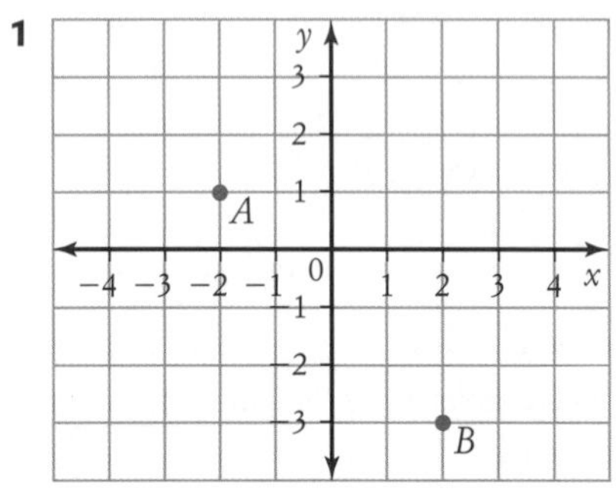

$D(-1, 2), E(-4, -2)$

2

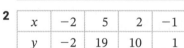

x	-2	5	2	-1
y	-2	19	10	1

3 a $y = x + 4$ **b** $y = 5x - 4$

Part D

1 y-coordinate
2 a A quarter of the number plane bordered by the x- and y-axes

b
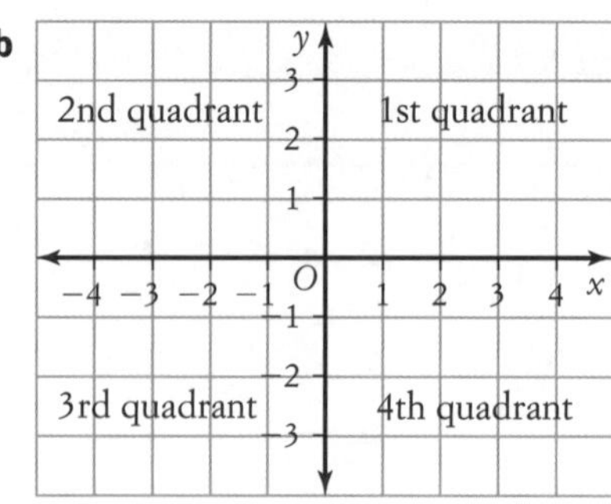

3 $(3, -2)$ **4** On the y-axis
5 a 11 **b** $y = 4x + 7$
6 positive, negative

Graphing tables of values PAGE 168

Part A

1 2 **2** $16\frac{2}{3}\%$
3 $90 = 2 \times 3^2 \times 5$ **4** $t = 6$
5 $14\ \text{cm}^2$ **6** 5.58 p.m.
7 23.5 **8** $x = 6$

Part B

1 $A(-1, -3), Q(2, -2)$
2
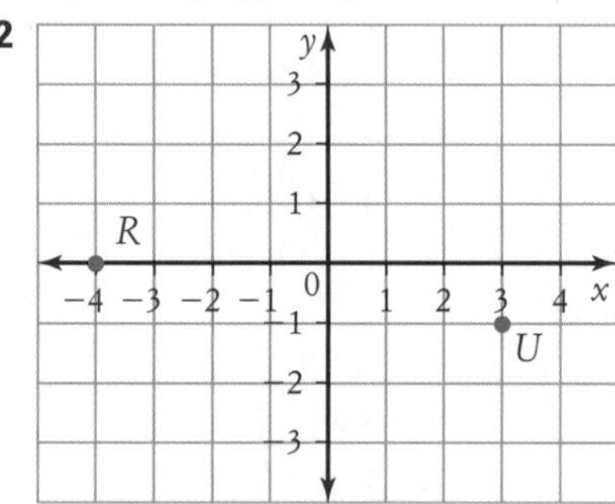

3 $(-3, 1)$ or any ordered pair with a negative x-coordinate and positive y-coordinate
4

x	-1	0	1	2
y	-5	-2	1	4

5 a $y = 4x$ **b** $y = 2x - 2$

Part C

1 a 19, 22 **b** $T = 3n + 4$
c 64 **2a** 7, 9, 11
b $y = 2x + 1$ **c** 31

9780170454520

2 d and **3**

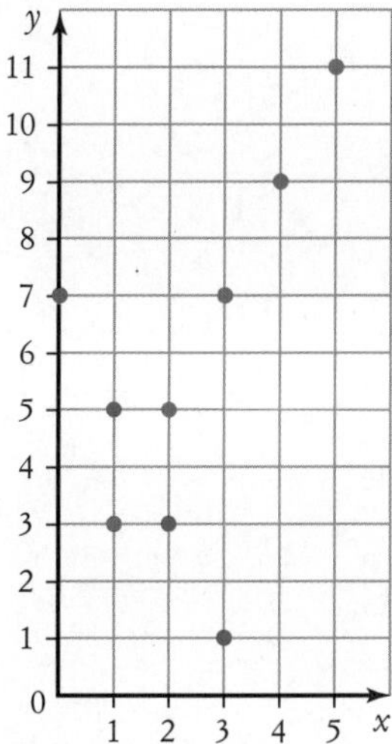

Part D

1 **a** 3rd quadrant **b** x-axis

2 **a** 4, 7, 10, 13, 16 **b** increasing by 3

c The y-values increase by 3, so the formula is of the form $y = 3x$ _____. Testing one ordered pair such as (2, 7): $3 \times 2 + 1 = 7$, so the rule is $y = 3x + 1$.

d 37 **3** origin

4 $y = x$

Graphing linear equations

PAGE 170

Part A

1 **a** 16 **b** 1

2 4 **3** $3a + 3$

4 22 m^2 **5** 8 times

6 $x = 11$ or 2 **7** 95°

Part B

1 **a** 5 **b** $y = 3x - 1$

c see 4 red dots

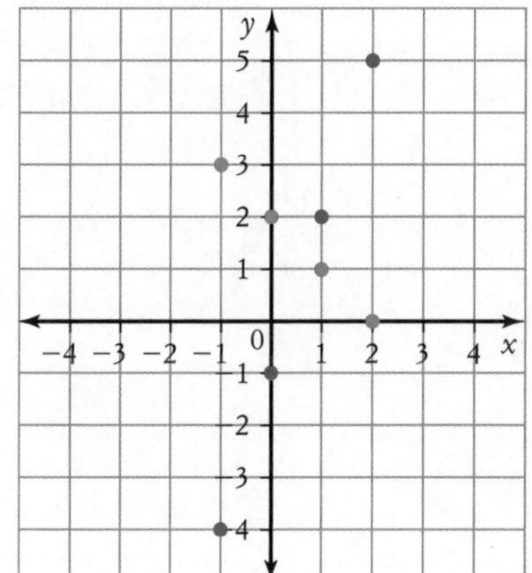

2 **a** 3, 2, 1, 0 **b** See 4 green dots on **1c**

3 **a** 44 **b** $T = 5n + 9$

c 509

Part C

1

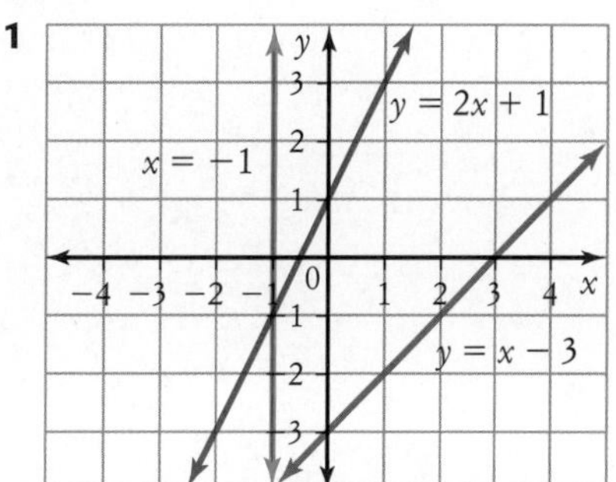

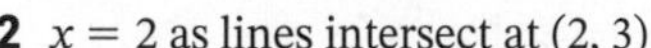

2 $x = 2$ as lines intersect at (2, 3)

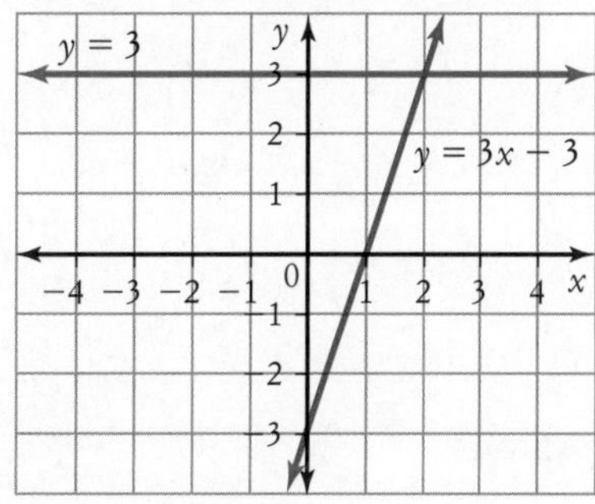

Part D

1 (1, 0)

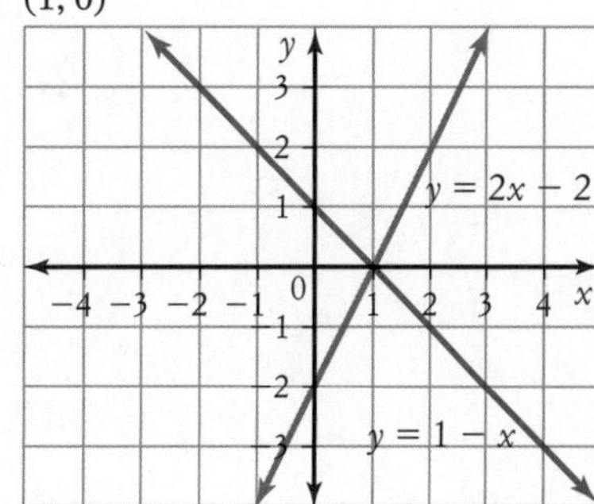

2 **a** $y = 2$ **b** $x = -3$

3 (3, 5) satisfies the equation $y = 3x - 4$. When $x = 3$, $y = 3 \times 3 - 4 = 5$. So (3, 5) lies on the line $y = 3x - 4$.